Einstein

for Anyone: A Quick Read

Second Revised Edition

David R. Topper

Vernon Series on the History of Science

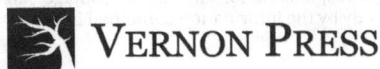
VERNON PRESS

Copyright © 2016 Vernon Press, an imprint of Vernon Art and Science Inc, on behalf of the author.

All rights reserved. No part of this publication may be reproduced, stored in a retrieval system, or transmitted in any form or by any means, electronic, mechanical, photocopying, recording, or otherwise, without the prior permission of Vernon Art and Science Inc.

www.vernonpress.com

In the Americas:
Vernon Press
1000 N West Street,
Suite 1200, Wilmington,
Delaware 19801
United States

In the rest of the world:
Vernon Press
C/Sancti Espiritu 17,
Malaga, 29006
Spain

Vernon Series on the History of Science

Library of Congress Control Number: 2016914467

ISBN: 978-1-62273-199-2

Product and company names mentioned in this work are the trademarks of their respective owners. Photographs used under permission by Art Resource Inc. While every care has been taken in preparing this work, neither the author nor Vernon Art and Science Inc. may be held responsible for any loss or damage caused or alleged to be caused directly or indirectly by the information contained in it.

for Alexis

Alexis loves gravity:
Perched in chair on high,
She laughs gleefully,
Droppin' all that's nigh.

Table of Contents

Foreword	1
Why this book?	1
I. The Smile	3
II. Love	15
III. Race	33
IV. Chutzpah	53
V. The Hair	93
What are you missing?	103
Bibliography	105
EndNotes	111

List of Figures

Figure 1
My conceptual images, at about the age of five, of how the world could be round.   54

Figure 2
The 3-D analogy of Einstein's 4-D model of the universe as curved and finite.   57

Figure 3
My reconstruction of Einstein's thought experiment about traveling at the speed of light.   58

Figure 4
Newton's drawing of his thought experiment of a falling projectile going into orbit around the earth.   64

Figure 5
Einstein's 1907 thought experiment on the identity of gravity and acceleration.   66

Figure 6
The 3-D analogy of Einstein's 4-D model of gravity.   68

Photos and Credits

Photo 1a
Einstein's school class photo of 1889. He is in the front row, 3rd from the right. Credit: bpk, Berlin / Art Resource, NY.   3

Photo 1b
Close-up of Einstein's smile in Photo 1. Credit: bpk, Berlin / Art Resource, NY   4

Photo 2
Einstein in 1931 at a reception in the German Chancellery in Berlin. Credit: bpk, Berlin / Art Resource, NY. Erich Salomon, photographer.   6

Photo 3
Einstein's 1896 Aarau class photo. He is far left in the front row. Credit: Art Resource, NY.   13

Photo 4
Einstein circa 1930. Credit: Universal Images Group / Art Resource, NY.   94

Photo 5
Einstein circa 1940, with a halo of hair. Credit: Culver Pictures / The Art Archive at Art Resource, NY.   96

Foreword

David Topper has provided in this book a fascinating introduction to Einstein's career and scientific achievements. I would like to present here an example of how a major world event saved his reputation.

—

On June 28, 1914, the Archduke Francis Ferdinand of Austria-Hungary was assassinated in Sarajevo by a Serbian nationalist, Gavrilo Princip. This event was the immediate cause of World War I. As we might say today, it was like the flapping of a butterfly's wings, which led to a four-year hurricane that devastated Europe.

It also had an indirect (one might say beneficial) effect on the fate of Albert Einstein's general theory of relativity. A German astronomical expedition led by Erwin Findlay Freundlich went to the Crimea peninsula in Russia, the best place to observe the solar eclipse scheduled for August 1, 1914. They wanted to test Einstein's prediction that starlight will be deflected by an angle of 0.87 seconds near the edge of the sun. But on August 21, 1914, Germany declared war on Russia, and the Russians therefore arrested the German astronomers as enemy aliens, preventing them from making observations. Had the astronomers done so with sufficient accuracy, they would have found that the deflection is actually 1.74 seconds — twice as much as the prediction from Einstein's theory

Einstein later revised his general theory, and predicted on November 18, 1915 that the deflection should be 1.74 seconds. Another expedition led by British astronomer Arthur S. Eddington went to the island of Principe (in the Gulf of Guinea off the west coast of central Africa) to observe a solar eclipse that was to occur on May 29, 1919. Fortunately for science, the war had ended on

November 11, 1918, so such observations could be made without risk of military interference.

Eddington analyzed the observations and announced on November 6, 1919 that Einstein's (new) prediction had been confirmed. The result was enormous publicity for Einstein and his theory, starting the next day when the *Times* of London proclaimed a revolution in science started by "one of the greatest achievements in human thought."

Einstein quickly became the most famous scientist in the world and remained so long after his death; he was named "person of the century" by *Time* magazine in 1999. But suppose Princip had missed his one chance to kill the Archduke – someone might have bumped into him as he was about to fire his gun – then Einstein would have been in the embarrassing position of having to revise his prediction after it had been refuted, and probably would not have gained such a high reputation.

(For further details, see my book *Making 20th Century Science: How Theories Became Knowledge* [Oxford University Press, 2015, pp. 488-490.])

Stephen Brush

Why this book?

When Albert Einstein became a world-famous celebrity in the 1920s, there followed an avalanche of books about him and his theory, a phenomenon that continues today. Most of the books have gone out-of-print, many are out-of-date – and so it is not superfluous to put another book on Einstein into circulation. Especially since there is a need for one that fills this gap: a very short book on Einstein that gives a brief but up-to-date story of his life and thought, with a short and simple explanation of what he contributed to 20th century physics. I hope my little book, which can be read in a sitting or two, is up to the task.

On the format: this book is written in a casual style, and I sometimes use contemporary acronyms, such as FYI ("for your interest") or BTW ("by the way"). Also, now and then, I like to begin an idea with the phrase "knowing what I know," when I wish to speculate on something based on my prior knowledge. For this I will use the acronym KWIK, which (if you wish to pronounce it) would be "quick."

(Occasionally, I put some information in a paragraph in parenthesis, like this one. I do this when the material is too interesting and relevant to be just dumped into an Endnote in the back of the book, but it still is not directly germane to the topic in the text. As such, it is a more like a sidebar, and you may just read around it. Note: it is too late to skip this one.)

I very much wish to thank Barbara Wolff (retired) of the Einstein Archives, Jerusalem, for her close reading of the first edition of this book. Her valuable suggestions and corrections were incorporated into this revised edition. Nonetheless, I am solely responsible for all errors herein.

I. The Smile

My first recollection of hearing the name "Einstein" was around the age of ten or eleven. I am putting his name in quotes because I heard it as an eponym, as in, "Well, you are no Einstein." I thought this was a compliment, for I assumed it meant something similar to, "Well, you are no Frankenstein." I soon found out that this "Einstein" was really the smartest person ever, and for some reason I thought that he was dead. Little did I know that he was still alive. Moreover, later in life I learned that if I had sent him a letter at the time, I may have received a reply. Einstein had a thing about writing to children. So, I now could have a signed letter from "A. Einstein," for me to cherish. Too bad.

Photo 1a
Einstein's school class photo of 1889. He is in the front row, 3rd from the right. Credit: bpk, Berlin / Art Resource, NY.

When the person Albert Einstein was about the same age, he had his class photo taken in school. This picture (Photo **1a**) is one of the first pictures we have of him, and is one of my favorites.[1] It shows his 1889 all-boys class of 52 students lined up in five rows outside their school. Einstein is in the front row, 3rd from the right, and clearly smaller than (perhaps the smallest of) all the others.

The unique and utterly fascinating thing about this picture is the simple fact that of *all* the boys looking grimly at the camera, little Albert is the *only* one with a smile on his face.[2] All 51 others, with hands at their sides, appear stern, anxious, intimidated, sulky, or scared; whereas Einstein, with hands behind his back, has a cute, little, slightly impish smirk on his face – unquestionably, a look that any parent would love. Just compare the detail of him with the boys to his sides (Photo **1b**): the contrast, indeed, is at once stunning and amusing.

Photo 1b
Close-up of Einstein's smile in Photo 1. Credit: bpk, Berlin / Art Resource, NY

I. The Smile

Right here, in this astounding image (a mere class photo) is the visual manifestation of the laid-back contrarian that he would become throughout his life. In 60s jargon, he would follow the beat of a different drummer. In this one picture, KWIK, Einstein's whole life (sort of) flashes forward before me.

Let's briefly jump ahead to 1931, with Einstein the celebrity at a reception in the German Chancellery in Berlin (Photo 2). From the left there are: Max Planck (the famous physicist), Ramsay MacDonald (British Prime Minister), Einstein, Hermann Schmitz (on Einstein's immediate left [FYI: Schmitz was from I.G. Farben, the chemical company that would become notorious for its role in developing Zyklon B used in the gas chambers in the Extermination Camps, and for which Herr Schmitz would spend time in prison for Nazi war crimes after World War II]), and Hermann Dietrich (German Finance Minister). While we are flashing forward to the War, it may be of interest to note that one of Planck's sons, Erwin – who was also present at this formal affair but is not in this picture – was executed by the Nazis as part of the plot to assassinate Hitler on July 20, 1944. BTW: such depressing matters have a way of just popping up in 20th century German history.

Photo 2
Einstein in 1931 at a reception in the German Chancellery in Berlin.
Credit: bpk, Berlin / Art Resource, NY. Erich Salomon, photographer.

Returning to the picture, itself: I have no idea why these five men were seated together or what they are talking about. There are several extant pictures of this table-talk scene, which were taken by the pioneering photojournalist photographer, Erich Salomon. I have chosen this one because it captures an animated Einstein speaking to the British Prime Minister. It is a captivating image clearly displaying Einstein's alert and smiling face along with his expressive right-hand gesture, all in stark contrast to the serious, stern, and solemn visages of the other four. "Come on, guys – lighten up!" – I want to say with Einstein. Or, put differently: What's there not to like about this Einstein fellow trying to cheer up a much too formal table? It is clear why I juxtaposed this 1931 picture with the smiling boy in school.

But there is more. Take one last look at Photo **2**, for it will never appear the same to you after I tell you this. While researching further into the story of this photograph, the joviality of this image was abruptly crushed for me when I discovered that the photographer Salomon died in 1944 at Auschwitz, the infamous Camp, which was supplied with chemicals from I.G. Farben. The result is a photograph deeply laden with meaning. In short, I can never again see this picture with that initial innocence. As I said, depressing matters keep popping up... More on this later in the chapter on Race.

As reported by those who knew him, Einstein was modest and unpretentious, without an iota conceit or arrogance, treating all people in the same manner, independently of class or rank. He spoke the same way to a president as to a janitor. [3] He also had a hearty laugh, with a child-like twinkle in his eye. OK, all this may be a bit of an exaggeration (sounding more like Santa Claus), but variations of these traits are persistently repeated among those who knew him and reminisce about his personality. He really was a down-to-earth guy. For example, he refused to travel first-class; even when he was sent first-class tickets he sat in the third-class, driving the ticket-takers crazy.

Back to the student years. What do we know about his early life that might help us to read more deeply into Photo **1**? Best to begin at birth.

Albert Einstein (1879-1955) was born in a small town (Ulm) on the Danube River in southwestern Germany to unobservant Jewish parents. Although the town today boasts of his birth, while he was still an infant the family moved to Munich, where he spent his formative years. His mother, Pauline, had a deep commitment to music, and she tried to instill that affection into her young son by forcing violin lessons on him. A love of music eventually sunk into his psyche around his transition to the teenage years, and Albert carried that commitment throughout his life. He exhibited his love of music by packing his violin on trips. Serious music, to him, was confined to the works of the "classical" period of what is called classical music, especially that of Mozart and Haydn, but for him we should also dip back into J. S. Bach.

His father, Hermann, was a businessman who could have made a lot of money at the time because he was in the electrical business (motors and dynamos, for example), which was to the late-19th century what computer high-tech stuff was to the late-20th century. But, just as the "dot.com" boom and bust resulted in some winners and many losers, most who made the effort in the electrical business did not achieve success. Hermann's business went bust.

Albert's sister, Marie (called Maja), was born when he was age two, and she was his only sibling. Maja, in a short memoir written in the early-1920s, is a crucial source of information about her brother's childhood, which is important because there are many myths circulating through the media and beyond about Einstein's youth. Today, special interest groups wish to embrace Einstein as the poster boy for their various causes. Nonetheless, Einstein was not a slow learner, a vegetarian, left-handed, or any of a range of idiosyncrasies that you will find in special-group websites on the Internet testifying that Einstein was one-of-them. Although his parents tutored him for his first year of school, he also was not "home schooled," for he continued through the German school system until the age of fifteen when he dropped out before graduating.[4] Contrary to another myth, Maja reports that her brother was "a precocious young man" who had a "remarkable power of concentration," such that he could "lose himself...completely in a problem."[5] KWIK, for Einstein the scientist, this youthful behavior was clearly repeated, like a leitmotif, throughout his scientific life.

It is true that Albert detested the rigidity of the German way of teaching, but he still got good grades. Yet he did not hide his feelings about the oppressive atmosphere of the classroom, so that one teacher went so far as to tell Albert's parents that their son set a poor example by his hostility. This may cast some light on the smile on his face in our photo, for it surely reveals the contrarian attitude on social matters that he would display throughout his life. One obvious example: think of his lack of decorum in the grooming of his hair, which began around 1930. (More on this in Chapter **V**.)

I. The Smile

An example of nonconformity of a different kind took place in his pre-teen years when he became extremely religious and admonished his anti-religious parents for not following the rules of Orthodox Judaism. This personal obsession lasted for a few years, at the consternation of Hermann and Pauline, only to disappear right before he would have been Bar Mitzvah. In his very brief autobiography, written in 1947, he says that the reason for the quick change was his discovery of science and the accompanying realization that the Bible was untrue. The result was an intellectual and emotional transformation. He viewed the religious outlook as subjective and solipsistic, whereas the scientific viewpoint was a route to objectivity and a liberation from what he called "the merely personal." He put it this way: "Beyond the self there is the vast world, which exists independently of human beings, and that stands before us like a great, eternal riddle, at least partially accessible to our inspection and thinking." This statement – as will be seen – acted as a maxim for his scientific endeavors to the end of his life.

But this is not the full story of his transformation: he added a socio-political element that is rather startling and remarkable for someone around age twelve or thirteen. He said he came to realize that "youth is intentionally being deceived by the state through lies" and that therefore a "mistrust of every kind of authority grew out of this experience." [6] These are profound and troubling views for someone at an age where most boys are more obsessed with sports and girls. Does this give us a hint at the meaning of the smile in Photo 1? Maybe not, since by age twelve we are a year or two beyond when the picture was taken. At least, however, we see the seeds of the unconventional citizen later in life.

Nonetheless, as we continue to pursue the question of the roots of his maverick ways, we find two episodes at about age fifteen or sixteen of interest. Both were triggered by the collapse of his father's business, and the need for the family to move from Munich to the town of Pavia in northern Italy just south of Milan, where his father's brother had a more successful business. Since Albert was still in high school, he was placed in the home of family members in Munich, while his parents and sister went on to Italy without him. Alone and feeling abandoned, he sunk into a deep depression and decided to leave school, hence his dropping-out. But he still had enough levelheadedness to obtain a letter from his math teacher saying that he completed that part of the curriculum. This was the first episode.

The other episode, however, might not have seemed very levelheaded at the time. After arriving in Italy, he applied to the German government to renounce his citizenship, making him a stateless person thereafter. Some scholars believe that in order to trigger such an extreme act, something almost elemental about German society had deeply troubled Einstein.[7] We know he had major misgivings about the militaristic features of German society as expressed in the educational system. Or, was it a reaction to his father's loss of his livelihood, and the need to leave the country? His sister, Maja, however, had a simple answer: he was avoiding being drafted into the military. [8]

Accordingly, as a high school dropout, Albert arrived at his parents' residence in Italy, much to their surprise and surely their chagrin. We have no documentation about the inevitable confrontation between him and his parents, but we can be sure that there was a dispute around the question of what he was going to do with the rest of his life. We, of course, know the answer, in the long run. But even in the short run, there was some hope.

Let's return to that letter in Albert's pocket when he left Munich, and back up a few years to the non-Bar Mitzvah around age twelve to thirteen. The unperformed religious transformative rite was, indeed, replaced by a different revelation – as mentioned, he developed a zeal for science, and in particular the logical rigor of mathematical reasoning. Specifically, he was given a primer on geometry and he devoured it, even trying to prove some theorems before he read the proofs in the book. The logical way mathematical reasoning produced eternal proofs had a deep psychological impact on this young man, so much so that even when writing his autobiography around the age of sixty-eight, he referred to this early textbook as the "holy geometry book."[9] How revealing the metaphor was: especially when we realize that he was reading Euclid instead of *Torah* (*The Bible*), the original "holy" book. He went on to teach himself calculus and other higher mathematics, so that by the time of his dropping-out of school, he was well-grounded in the mathematics required for graduation and beyond. Hence, the letter in his pocket.

Albert's father had plans for his son to be an engineer. This is no surprise, since he was in the electrical business, which he (correctly) believed was the wave of the future. In particular, he wanted his son to enroll in the Swiss Polytechnic Institute in Zürich, one of the best schools in Europe. As luck (fate?) would have it, under special conditions a completed high school diploma was not necessarily required for enrollment in the Poly; instead, there were a series of rigorous exams administered by the Institute. It is possible that the letter from the math teacher was a factor in placing him in the special category.

So, in the fall of 1895 he took the entrance exams – but flunked them. There was, however, a silver lining to this incident. He did so well on the science and math parts (no shock here) that the Institute's director recommended that he spend a year doing some remedial studying. After all, he was applying to the Institute a year or two early for his age, since the regular age of admission was about eighteen years old.[10]

Einstein spent this year at the Kanton Schule in the town of Aarau, just west of Zürich. The curriculum was based on the ideas of the great Swiss educator, J. H. Pestalozzi, who (among other things) emphasized using visual materials as well as written texts as educational tools, and especially stressed direct student-teacher interaction.[11] For Einstein, it was a delightful and memorable year: he enjoyed learning in a formal setting for the first time in his life, and as a bonus, he had the first love affair of his life. He lodged with a family of one of the teachers, Jost Winteler, and the love object was one of the daughters, Marie.

(For future reference: another daughter [Anna] later married one of Einstein's best friends [Michele Angelo Besso], and Einstein's sister married one of the sons [Paul].)

It was sometime during that year of motivated learning that he came up with what would be his first great experiment in his head, what we call a "thought experiment." This idea involved moving in space at the speed of light; essentially it was based on this question: What would the world look like if we rode on a beam of light? Perhaps the Pestalozzi emphasis on visualizing played a role here? We will certainly come back to this in Chapter **IV**. For now, listen to the following remark about the school in Aarau that Einstein wrote sixty years later: "It made an unforgettable impression on me, thanks to its liberal spirit and the simple earnestness of the teachers who based themselves on no external authority." [12] Ah ha, "no external authority": such progressive and open-minded thinking was guaranteed to have an impact on Einstein who, as quoted, believed that "youth is intentionally being deceived by the state through lies" and that therefore a "mistrust of every kind of authority grew out of this experience." This Swiss Kanton Schule was obviously nothing like the German schooling he had experienced. No wonder he graduated in the fall of 1896 with good grades.

The year at Aarau proved fruitful. Einstein's admittance to the Swiss Polytechnic was based on his grades at Aarau, and although his father wanted him to study to become an engineer, he enrolled in physics and mathematics with the aim of becoming a teacher.

This chapter began with his 1889 class photo; it ends with another one, seven years later (1896). Photo **3** shows the small group of students in the Aarau class, with Einstein plainly seen in the front row seated on our left. No smiling faces here, by him or any other of the ten students. Yet there is uniqueness in Einstein's countenance. Slightly slouching as he sits back relaxed and crossed-legged, his tie loose and collar open, he displays an air of self-assuredness.[13] In contrast, his nine fellow students assume stiffer and more posturing poses; ties and collars tightly done,

I. The Smile

there is even a hand (Napoleon-like) thrust into a coat. Either they look at the camera or slightly to the side at some nearby objects. Einstein, however, gazes far off into the distance – beyond the mundane shortsightedness of his fellow students.

Photo 3
Einstein's 1896 Aarau class photo. He is far left in the front row. Credit: Art Resource, NY.

This picture visually brings to mind what he once reported: while the other students in Aarau filled their spare time by swigging copious quantities of beer, he instead drank from a different trough – diligently reading *The Critique of Pure Reason*, by Immanuel Kant. [14] And that was nothing to smile about.

II. Love

When I was an undergraduate student studying physics and mathematics in the early 1960s, there was only one girl in our advanced physics classes. As I learned later, nothing much had changed since the days of Einstein; there was just one girl in his classes at the Poly too. She was a Serbian, Mileva Marić, and he fell in love with her. (Note: I did not copy Einstein in this action. Of course, there are many other ways I am not like him, but let's change the subject.) It seems that Einstein was the only guy attracted to her.

Mileva was a plain girl with a slight limp. There are sources calling her ugly, but these are surely hostile remarks coming later from those who had a grudge against her. It is true that in middle-age, she did not take care of her grooming and the pictures that we have are of a heavy, unattractive, and clearly unhappy woman. But the photos of her when she met Einstein show an alert, brightly looking, pretty face. She certainly had to be smart and gutsy to get to where she was as a woman in *fin de siècle* Europe.

(BTW: I have been thinking recently about the how what we know and feel about people play a key role in our perception of them. I was made bluntly aware of this when I came across a large poster of Malala Yousafzai, whose face has not fully recovered from the damage done by the bullet fired by a would-be Taliban assassin on October 9, 2012 – a bullet that passed right through her head. I found myself staring transfixed at the picture and muttering under my breath: "Malala is one of the most beautiful faces in the world.")

Einstein was also close friends with an Italian Jew, Michele Angelo Besso; he is the person mentioned in Chapter **I** who married the other daughter (Anna) of the Winteler family at Aarau. Einstein, recall, fell in love with Marie Winteler; and since *Love* is the topic of this chapter, we will look back at this affair before continuing with his life at the Poly.

Regrettably, our knowledge of his relationship with Marie is minimal and fragmented, but what we know is very tantalizing. Starting in the fall of 1895, the relationship grew over the year – quite intense for Marie, and fervent for Einstein too. In a letter to Marie in the spring he called her his "sweetheart," "my dear little sunshine," and his "naughty little angel." You get the drift... yes, this is Einstein, but as a 19th century Romantic teenager. Want more? "You mean more to my soul than the whole world did before [I met you]." [15] Gag?

The relationship continued when he moved to Zürich and the Poly, but after he met Mileva he lost interest in seriously continuing the liaison with Marie – although, apparently, he still sent her his dirty laundry to wash and iron. What he viewed as a sociable act in keeping with his Victorian view of woman, she interpreted as a domestic act foretelling a permanent relationship. When he did not respond to her letters, she interpreted it as a rebuff. This is my nuanced version of the story. The short and severe report of some other historians is that he used Marie and broke her heart. [16]

It is true that Einstein did not then, nor ever, display anything approaching a late-20th century feminist worldview; few men did at that time. But a teenage boy (he was seventeen years old) dumping a girlfriend (she was nineteen) does not a misogynist make. Let's get real: it was teenage love. Perhaps the most interesting insightful fragment of information we may glean from this affair is found in a letter he wrote to Marie's mother in May 1897, long after he had broken off the relationship. Probably written out of guilt, or at least some remorse, he spoke of a "pain" which he inflicted on "the dear girl through my thoughtlessness and ignorance of her delicate nature." He could (should?) have left the matter there, with a further *mea culpa* and signature. But not Einstein. Instead, in the very next sentence, he went off on a tangent about the travails of his work in science, as if this excused

II. Love

his insensitivity in social relations. Moreover, he put it forth in highfalutin language about his "strenuous work" in probing into "God's Nature" as a way of weathering the "storms of life." [17] Actually he was right, but not in sense that he was (as we now know) really looking into the laws of nature; rather he was admitting that when it came to making choices in life, physics always came first. Contrary to what he told Marie, he actually gave over his soul to physics. That was his Faustian bargain. Physics was the real *Love* of his life, which is the theme of this chapter, if not this entire book.

So Marie was displaced by Mileva, who was now the focus of his affection. The relationship became quite intensive very rapidly, such that Mileva left the Poly to take courses instead at Heidelberg University, 150 miles north of Zürich. Yet they continued the relationship by mail.

One letter from Mileva is interesting in two ways. She discussed a physics lecture on kinetic gas theory. "Oh, it was really neat," she reported, referring to the way the professor calculated colliding molecules. [18] This reveals that the passion between these two students was not only of the flesh. She too had a thing for physics. The other thing of interest is that the professor was Philipp Lenard (1905 Nobel winner) who will later enter Einstein's life in a very sinister way (in Chapter **III**). Einstein eventually coaxed her back to Zürich and they were again a couple.

Their surviving correspondence reveals a repeat of the Romantic rhetoric about love that Einstein displayed with Marie, but with an additional love – the love of physics that he and Mileva shared. "How was I able to live alone before [I met] my little everything? Without you I lack self-confidence, pleasure in work, pleasure in living – in short, without you my life is no life." [19] He could have cut and pasted this schmaltz from a letter to Marie – but Mileva was different in that he could juxtapose such cornball stuff with, say, a discussion of something as sexy as the latent heat of metals. [20]

Mileva was also different in another way. Whereas Marie was conventional and bourgeois; Mileva was an Orthodox Serbian, and this gave her an exotic flair in Einstein's mind. In his letters he called her Dolly and himself Johnny, and often spoke of her as a little witch. He contrasted their free-spirited life together with his home life, which was "narrow-minded and philistine." [21] He liked to use the word "philistine" to label the world they were rejecting, and it appears often in the letters.[22] With her he fantasized living an alternative bohemian life replete with his two loves. "I cannot wait to have you again, my all, my little beast, my street urchin, my little brat," and in the same letter he reports that a book on the physics of gases by Ludwig Boltzmann that he was reading is "magnificent." [23]

Completing his four-year degree at the Poly, Einstein graduated in the summer 1900 with respectable marks. Nonetheless, he could not find a job. It was customary for the teachers at the Poly to find employment for the good students, but no one made the effort for him. We do not have to look far for some answers as to why. As a student, he exhibited his contrarian behaviour overtly: he often cut classes, asked questions teacher found provocative, and did not follow social protocol.

Unlike most class-cutting students, however, Einstein was not goofing-off. Instead, he was often in the laboratories performing experiments. This may seem strange, since the common image of Einstein as a scientist is his making abstruse calculations on a blackboard rather than tinkering with some gizmo in a lab. But recall that his father's business was electrical motors and dynamos and other such high-tech late-19th century gear, and the son liked to spend time learning hands-on in the factory; in addition, the labs at the Poly were some of the best-equipped in Europe. Regarding the classes he missed? He had a very good friend, Marcel Grossmann, who took copious and lucid notes, and they were sufficient for him to study in order to pass the courses; after all, we are talking about Einstein!

II. Love

When not in the labs he, alternately, was in the Poly's Library reading texts not on the curriculum: contemporary works by James Clerk Maxwell, Ludwig Boltzmann, Hermann von Helmholtz, Ernst Mach, and others. (It is hard to believe, from what we know today, that at least Maxwell's electrodynamics was not being taught.) Einstein surely did not curry favour with his profs when asking why this material was not taught in the courses. Finally, his not using the proper salutations when addressing his teachers was seen by them as further hostile behaviour. That's why – no job.

He eventually found some part-time teaching, and earned some income doing private tutoring in physics and mathematics. Mileva, in a letter to a friend, gave two reasons for his lack of secure work: "My darling has a very wicked tongue and on top of it he is a Jew." [24] She was obviously aware of the latent anti-Semitism that was endemic in *fin de siècle* European culture. (More on that in the next chapter.) In the meantime, Einstein began working on his PhD at the University of Zürich, a task that would add pressure on his endeavour to earn a living. Moreover, if that was not enough, Mileva became pregnant. Being an unwed mother and the companion of Einstein at this time would have jeopardized his chance for employment. In the end, Mileva returned to her family to give birth to the child, alone.

We do not know if the possibility of them breaking up – essentially with Einstein dumping her and being what today is called a "deadbeat dad" – was ever considered. Apparently not: they were much in love at the time; and, as I have argued, Einstein was not as malicious as some historians try to portray him, at least not at this time. In fact, in a letter written to Mileva, probably in May 1901, he told her to "be of good cheer, love, and don't fret. After all, I am not leaving you and I'll bring everything to a happy conclusion." The editors of the *Einstein Papers* surmise that the original plan was for Mileva to return to Zürich with her child. But that never happened. The baby, a girl they named Lieserl, was born probably in January 1902, and she stayed with Mileva's parents in what is present-day Serbia (at the time it was part of the Austro-Hungarian Empire). As a result, Einstein never saw his first child.

Apparently, she was given up for adoption, for Mileva returned to Zürich without their child. Historians did not even know about the existence of Lieserl until the mid-1980s when her name appeared in a series of letters that were opened to the public for the first time. Subsequently, efforts to track down what happened to her were extensive but, so far, fruitless.[25]

As this drama unfolded, Grossmann came to Einstein's rescue once more, by getting him an interview for a job at the Swiss Patent office in Bern. While Einstein was waiting for confirmation of the job, he placed an ad in the Bern newspaper that he would teach "physics for three francs an hour." He had two responses: Maurice Solovine and Conrad Habicht. The friendship with Solovine lasted his entire life and is chronicled in letters exchanged between them for almost fifty years, the last one written near Einstein's death. [26] Habicht went on to obtain a PhD in mathematics and become a high school teacher. When Einstein got settled in the patent office job, he recommended his friend Besso, who was hired. So now Einstein had another sounding board, as he thought deeply about the fundamentals of physics. The trio met weekly for two to three years, reading books in science and philosophy and calling themselves the Olympia Academy. Some of their readings and discussions were seminal in stimulating Einstein's thinking about space and time. Besso seemingly was the most important help, for when he published his first paper on the theory of relativity in 1905, Einstein only thanked one person – Besso.

It is of more than passing interest to pause here to remember Mileva at this time, since in the courting love letters he spoke of them working together on their mutual passion about physics. In what may be the last letter before the birth of Lieserl, he wrote: "When you are my dear little wife, we will zealously do scientific work together, so as not to become old philistines, right?" The bohemian overtones surfaced in his endearments: "You must always remain my witch and my street urchin." [27]

One year later, in January 1903, they were married in a secular ceremony[28] witnessed by Solovine and Habicht, with no family member present. In May 1904, their first son, Hans Albert, was born. (Later in 1910, Eduard, called Tete, was born, their last child.) Regretfully, Mileva's collaboration with Einstein never materialized. Einstein withdrew into the all-male world of his three friends, while Mileva looked after the baby and the household. The Romantic whimsy of the courting years was over. [29]

Despite the long hours and six-day workweek, Einstein found spare time at the Patent Office and at home to ruminate on his ideas about physics – so much so that in 1905, between March and September, he put forth what came to be five landmark works in the history of physics. They were on the molecular theory of matter, the quantum theory of the atom, and his theory of relativity. (More on this in Chapter **IV**.) For this reason, the year 1905 is immortalized in the history of science.

At the time, however, the significance of his effort was not fully realized, although some physicists recognized the likely importance of what he had done. One important physicist who looked positively on it was Max Planck (in Photo **2**), who had put forth the concept of a quantum of energy in 1900. By 1906 he was lecturing on what he called the relative-theory. Another was Johannes Stark, editor of a yearbook in physics, who asked Einstein to write a summary article on relativity, which was published in 1907. (FYI: Stark received the Nobel Prize in 1919. Later, he, like Lenard, would turn on Einstein, with the rise of anti-Semitism.)

Yet it was not for another two years (1909) that he finally got his first salaried university appointment, at the University of Zürich, where he received his PhD in 1906. No longer a civil servant at the Patent Office, he was now a professor trying to continue his prodigious research work, while preparing lectures and dealing with students and faculty matters. In 1911 he accepted a position at the German University of Prague, a job that lasted only a year, since both he and Mileva were very unhappy with seemingly everything there – they found fault in the people, the social environment, the educational system for the boys, even the water. Einstein also bitched about the bureaucracy at the University – the endless paper work he called "ink-shitting." [30]

Nonetheless, while in Prague he was quite productive, especially for work on what became his theory of gravity. Also, despite his whining about Prague, he had a close circle of friends who were very important in his life. It was a group of Jewish intellectuals (one of whom was Franz Kafka, although he was not famous until after his death), who rekindled Einstein's connection with Judaism that he had broken-off in his preteen years.[31] (We will have more to say on this in the next chapter.)

After Prague, he went back to Zürich, and (of all places!) to the Swiss Polytechnic, his undergraduate school where he earned the distain of most teachers through his benign hostility. For example, his math teacher, Hermann Minkowski, called him a "lazy dog," although he later realized the significance of the modification of space and time in Einstein's relativity theory; and, in fact, in a famous lecture of 1908 Minkowski extended the theory by introducing the idea of a 4th dimension (see Chapter V). Sadly, when Einstein returned to the Poly, Minkowski was not on the faculty; he died (in his mid-40s) just as his famous lecture was being published. [32]

By 1913 Einstein was very well-known among theoretical physicists, due to his prodigious publications of highly original papers from 1905. In the spring of that year, physicists Planck and Walther Nernst from the University of Berlin came to Zürich to offer Einstein a prestigious research position with no formal teaching duties. (FYI: Planck received the Nobel Prize in 1918. Nernst was really a physical-chemist, who discovered what is called the third law of thermodynamics, for which he received the Nobel Prize in 1920.) In many ways, Berlin was the center of physics at the time, and so it was an offer difficult to refuse. Yet Einstein hesitated because he was now a Swiss citizen, and much more comfortable there than in the Germany he left and renounced nineteen years ago.

Nonetheless, he accepted the job: and, so, when he left the Poly in the spring of 1914, his formal teaching career of about five years was over. From anecdotal information, it appears that he was a very good teacher, explaining things slowly and clearly, and having a sense of humor and good rapport with the students. None of this is surprising, KWIK about the *only* student smiling in his class photo.

When the Einstein family moved to Berlin in the spring of 1914, the tension between Albert and Mileva seemed to echo the tension in the political climate of Europe. The latter led to World War I. The Einsteins' discontent had been stewing since around the miracle year of 1905. Albert was enamoured with physics as we know, and except for spending some time with his son, there was little time for Mileva and household matters. As seen, the promise of their working together and sharing a love of physics, as sketched by Albert in their courting years, never materialized. While he and his Olympia Academy pals discussed matters of philosophy and physics over tea and sausages, Mileva washed the dishes and kept Hans Albert occupied and out-of-the-way. Noteworthy is Solovine's report that she also listened attentively to the discussions but never took part in them.

Einstein later sometimes blamed the tension between them on Mileva: she was depressive, easily irritable, almost paranoid at times, he said. Maybe so, but who was at fault? There is no evidence that he was a model companion. Just recall that he still sent Marie Winteler his dirty laundry after he dumped her. Also Einstein was prone to simplify human behaviour by reducing it to genetics. Today we would speak of the role of DNA or of being hard-wired. For example, by the 1930s their younger son, Tete, was diagnosed with schizophrenia, and spent almost half of his adult life institutionalized. Since Mileva had a sister who was psychotic, he blamed Tete's problem on genetic factors coming from Mileva's family.

Yet, I believe, a perhaps more straightforward answer to Mileva's personality after the marriage was her guilt or shame in not raising Lieserl. Human losses can trigger long-time effects. Indeed, the adult Hans Albert reported that before he knew about this older sister, he was aware of his mother's persistent brooding over something that she called "very personal." [33]

Whatever the cause and whomever was to blame: as they grew further apart, the conflict became downright nasty. If we think of the saint-like photos of Einstein late in life, with the halo-like head of hair, and the deep penetrating eyes evoking an air of serenity and wisdom (Chapter **V**), it is hard to fathom him writing a hostile decree to Mileva in the summer of 1914 commanding her to perform all laundry (shades of Marie?), cooking, and housekeeping duties without any intimacy or any social interaction between them, unless he requested it. Essentially, she was to be merely the hired-help with a relationship lacking even an iota of companionship.[34] This was the nadir point to which their relationship had descended, after all the lovey-dovey of the student years.

But there was more at play when they moved to Berlin. As the two were drifting apart, Einstein become closer with a divorced cousin, Elsa Löwenthal, who was living in Berlin and whom he had known since childhood. Elsa was the opposite of Mileva in many ways. She was not an intellectual and had neither an interest in nor understanding of science. She was ordinary looking and chubby, with a delightful smile and an outgoing and very sociable personality that charmed many a guest at her home. Unlike Einstein, she liked being in groups, was able to make small-talk, and was a good judge of people. [35]

When he made trips to Berlin on scientific business, he would visit Elsa and her two daughters (Ilse and Margot) who lived with her. Quickly a bond developed between them, which went beyond mere cousinly friendliness. Mileva was aware of the friendship, and apparently, when viewed first-hand in the spring of 1914, she found the milieu unbearable, and so she left Berlin, taking the boys back to Zürich. Einstein was left alone in a new city with a prestigious job, but no wife and children. He missed the boys but not Mileva. As he wrote to a friend at the time: "I could not stand the wife any longer." [36]

In August, war began, and in Germany there was an outpouring of nationalistic jingoism that he found offensive and foolish. In particular, there was a patriotic declaration in support of German militarism circulating that ninety-three intellectuals (scientists, artists, physicians, theologians, and others) eventually signed. Many of his colleagues signed it, such as Planck, Nernst, and Fritz Haber (more on him in the next chapter). Einstein did not; instead, he signed an anti-nationalistic manifesto. This counter-document found only two other signatures. He wrote to a friend: "The international catastrophe has imposed a heavy burden upon me as an internationalist." Not since his renouncing his citizenship many years before was his antipathy to nationalism so overtly expressed. As he came to think more, and eventually write about political topics, we find him positioning himself as a pacifist, left-of-center, attracted to socialism with a corresponding antipathy to capitalism, and with a vision of a world government being the only solution to the stresses and strains inevitable within the nation-state system (more in Chapter V).

As German society was wallowing in the travails of war, Einstein escaped into the repose of mathematical physics. Since 1907, he was developing a further extension of the original 1905 relativity theory, trying to bring gravity into the equation, both figuratively and literally (Chapter IV). In November 1915 he brought to fruition the task he spent almost a decade working on – namely, his theory of gravity. He immediately went on to write a major summary paper of his theory in 1916, a popular book on relativity, and to top it all off – he created in 1917 a new theory of the entire universe based on relativity (Chapter IV). At the end of a lecture on how he created his theory of gravity, delivered at the University of Glasgow in the summer of 1933, he recalled "the years of anxious searching in the dark, with their intense longing, their alternations of confidence and exhaustion and the final emergence into the light" and concluded that "only those who have experienced it can understand it." This was not overblown rhetoric. It truly expressed what he went through over almost a decade starting in 1905.

Yet it was more. Let's read it again, and imagine that it is not about physics but about a lover, which brings out more empathically the erotic overtones. "Years of anxious searching in the dark, with their intense longing, their alternations of confidence and exhaustion and the final emergence into the light - only those who have experienced it can understand it." Ah, yes. It *is* about the supreme *Love* of his life – physics, of course.

Either due to the strain of this work or to some other causes, he became unwell about this time. For several years, he was plagued with various aliments: liver problems, stomach ulcers, and general malaise. Eventually he moved into an apartment in Elsa's building and was having meals with her. She was not demanding, giving him the space he demanded. Indeed, Elsa nursed him back to health with a special diet, peace and quiet, and probably most importantly, leaving him alone a lot.

World War I ended in November 1918 with Germany becoming a republic. In February 1919, Albert's divorce from Mileva was finalized and by June, he and Elsa were married. At the same time, his scientific life was also taking a major turn. His prediction that the sun would bend light from a star was confirmed by the Royal Society of London (Chapter **IV**), and the news was broadcast across the world as if the media were looking for something positive and uplifting to tell the world after four years of mud, mayhem, mustard gas, and death. Einstein had matched Newton in the pantheon of science and so the name Einstein no longer was a name familiar only to a small coterie of physicists. In a short time, "Einstein" was an eponym for smart, brainy, or genius – with the bonus that this Einstein fellow was a celebrity.

Initially he and Elsa relished being in the spotlight and smiling at the ever-flashing cameras. But the notoriety came at a price: a lack of privacy, and not being able to walk anywhere without people gawking at you. Predictably, in 1920s Germany, a Jewish scientist in the limelight brought out from the gutters anti-Semitic attacks. These attacks were sometimes camouflaged as critiques of relativity, although there were legitimate questions to be asked of the new theory (Chapter **III**). The stress of it all was too much for Elsa, who came down with various ailments including a "bladder infection" and "hemorrhages."[37] The stress was so great that they considered leaving Germany, for there were potential job offers in Zürich and Leiden. Philip Frank, Einstein's first major biographer, reports that when he met Einstein in 1921, as the theory of relativity was being condemned as anti-German, Einstein said he would be leaving the country in ten years. [38] (It turned out to be a dozen.) At one point in 1923, when a right-wing group made death threats against him, Einstein went to Leiden for six weeks, but he came back and resumed his usual duties.

Along with these trials and tribulations, Einstein was awarded the 1921 Nobel Prize for Physics, although it was not announced until 1922. (To be discussed in Chapter **IV**.) The notoriety led to all-expenses-paid trips; and thus much of the 1920s was spent traveling and lecturing in Europe and the United Kingdom, the United States, Japan, China, Palestine, and South America. Sometimes he went with Elsa, sometimes he was alone (although usually with his violin). The trips away from Europe were also an escape from the socio-political turmoil of the decade, particularly in Germany between the governing Weimar republic and the emerging far right that morphed into Nazism.

Another hazard of being a celebrity (some might call it a perk) was the adulation from various folk. Then, as today, this often entailed women flinging themselves at men's feet. This Einstein could not resist, and he had a series of liaisons during the decade. More or less, he was sowing his wild oats in his middle age. Elsa, at first, fought back, but eventually acquiesced when she realized that resistance was clearly futile. Albert was going to have his way with other women. For her it became a trade-off for the high-life of wining and dining with the elite at formal banquets she otherwise would not be invited to. After all, she probably enjoyed those occasion more than he did. She made small talk with high-society folk, while her bored husband scribbled equations on a napkin – unless, of course, there was an attractive woman next to him.

By the late-1920s Einstein was inundated with continuous correspondence that interfered with his scientific work. He needed a secretary. Elsa was leery of you-know-what and insisted on being privy to choosing the woman. In 1928, they hired Helen Dukas, a plain and proper woman who was no threat to Elsa.

(FYI: Dukas would remain with Einstein as his loyal secretary; after the death of Elsa, she became his platonic homemaker too – for the rest of his life. Dukas would become a trustee of his estate and receive $20,000 from his will.)

In January 1933, they (namely, Albert, Elsa, and Helen) were in California. It was their third annual winter sojourn to Caltech where the Nobel physicist, Robert Millikan, was hoping to get Einstein to move permanently. The previous fall, as Albert and Elsa were closing up their summer cottage to return to Berlin, he supposedly said: "Take a good look at it." She asked why. "Because you will never see it again."[39] He was proven right; they never did, for while Elsa and Albert were basking the sun of southern California, Adolf Hitler and the Nazi party were elected in Germany and so began a brutal dictatorship that was supposed to last 1000 years. (It only lasted a dozen years, but it was a dozen years too many.) A price was placed on Einstein's head, as the Gestapo ransacked their cottage and Berlin apartment, emptied his bank account, and burned his books. He had two choices for permanent positions in the USA: the west coast at Caltech, or the east coast at the new Institute for Advanced Study being built in Princeton, New Jersey. Einstein accepted the Institute's offer of a research position with no required teaching duties, and he remained there until his death in 1955.

Before taking up his post, they returned to Europe (but not Germany) in March 1933, staying in the coastal town of *Le Coq sur Mer* in Belgium and protected by security guards supplied by his friends, the King and Queen. He took a side trip to the UK to deliver talks in London, Oxford, and Glasgow. But probably the most important side trip was in May to Zürich, where he met his younger son, Eduard or Tete.

After the divorce, his relationships with his sons were strained. He made efforts to remain close to them through visits and letters, but there were periods of estrangement, possibly due to Mileva's bad-mouthing, but also to Einstein's rigidity (of all things!) in trying to control their behavior. He eventually reached a truce with Hans Albert as an adult, and they occasionally saw each other, especially after he took a position as an engineering professor at Berkeley, California.

But Eduard never left Europe. As seen, Einstein attributed his mental illness to hereditary factors from Mileva's family. As a result, the May 1933 trip to Zürich was bittersweet, for it was his last meeting with Eduard, and he probably knew this. Einstein continued with financial support, but after writing some unanswered letters to his son, he essentially cut off ties with him, since he viewed Eduard as hopelessly tied to his mother.[40] Was Einstein's seeming abandonment of Lieserl, without any apparent psychological baggage, a precursor of his behavior toward Eduard? Perhaps. All of this is in contrast to the apparent deep psychological scar on Mileva left by the loss of Lieserl, and maybe explains her deep devotion to Eduard to the end of her life.

Einstein returned to the USA and took up his new post at the Institute in October 1933. The next year Elsa's older daughter Ilse died in Paris, with Elsa present. The younger daughter, Margot, returned with her mother to Princeton. The Einsteins eventually purchased a modest house at 112 Mercer Street (it is still there with a private owner). Sadly, Elsa lived only a little over a year in the house that she refurbished, dying painfully of heart and kidney problems in December 1936. During her illness, Einstein was deeply distraught. Elsa's account is poignant: "He has been so upset by my illness. He wanders about like a lost soul. I never thought he loved me so much. And that comforts me." [41] Perhaps he loved her more than he realized.

In 1939, Einstein's sister, Maja, left her husband in Italy after anti-Semitic racial laws were passed, and she too moved into Mercer Street. When she was bedridden with terminal illness, Einstein read to her in the evenings. As a result, in his later years Einstein spent the domestic side of his life in the company of women. KWIK, we can say he probably was content, since he always liked to be around lots of women.

(FYI: Maja died in 1951, before him. As already noted, Dukas remained in the house after Einstein died in 1955, and until her death in 1982. Margot inherited the house and died in 1986.)

In the spring of 1955 Einstein was informed that Besso had died. Besso, recall, was a life-long friend, the only person thanked in the famous 1905 paper on relativity, who had married Anna Winteler, sister of Einstein's first love, Marie. In a letter of condolence to Besso's son and sister, dated March 21, 1955, less than a month before his own death, he wrote (in part): "What I admired in him as a human being is that he managed to live for so many years not only in peace but also in lasting harmony with a [single] woman – an undertaking in which I twice failed rather miserably."[42] Honestly put, Einstein admitted his shortfall in being committed to one woman at a time. The problem was that, despite his love of women, he liked physics more, much more. Maybe it is better stated this way: he *liked* women, but he *loved* physics.

Allow me to end this chapter with this example. Listen to the following sentence from a love letter to Mileva: "I am filled with such happiness and such joy that you absolutely must share in some of it." What, pray tell, was he talking about? It was not the birth of their child, for which we have no such reaction – although, to be fair, many letters are lost. After what you have seen in this chapter, I suspect you may have guessed that the topic of enthusiasm was physics – and so you would be correct. It was his reaction to reading a paper by Lenard (yes, him again). "I have just read a marvellous paper by Lenard on the production of cathode rays by ultraviolet light... [It is a] beautiful piece of work." [43] It turns out that this paper set the stage for one of Einstein's landmark papers of 1905 (see Chapter **IV**).

So, what's my point? I will set it as a question: How many people do you know who speak this way of, not a musical score or a work of literature, but of a physics paper? "I am filled with such happiness and such joy that you absolutely must share in some of it." Think about it: if such an expression were put forth about a work of art, few would blink an eye. So, why not physics? Frankly, it is a wonderful thing to see such an expression by a scientist, and to try to fathom how he sees beauty in a place were few, very few, do.

The point is that his love of physics was so supreme that it took precedence over everything else. Why not, when its beauty bestowed such happiness and joy?

Poets have died for less.

III. Race

In my last years of teaching, when I talked about Einstein in my classes, I was surprised how often a student would approach with the remark: "I didn't know that Einstein was Jewish." I was tempted to retort (but I did not): "And the Pope is Catholic." What I learned from the students' remark is that things some of us now take for granted are not so in the younger generation. All the more reason to write this book, I suppose. Moreover, this snooty remark by me (even if I did not say it) would be off-the-mark, anyway.

For the historian, Einstein's Jewish identity is a moving target. He waffled back and forth in his preteen years, and he didn't settle into a relatively stable identity until he was into his 40s. There is also some debate among those writing on this topic as to his allegiance to Judaism at different times in his life.

As to the title of this chapter: just as the word "love" had a dual meaning in Chapter II, so the word "race" has more than one connotation. Einstein's changing attitude toward Judaism, his so-called race, was often correlated to his exposure to anti-Semitism in Germany; later, after his move to the USA, he encountered (and was distress by) the discrimination against African-Americans – another form of racism.

As outlined in the previous chapters, Einstein's first real fling with Judaism was in those preteen years when he drove his parents batty trying to impose his orthodox views on the family. As reported in his autobiography, right before he was about to be Bar Mitzvah he abruptly abandoned this belief, as he immersed himself more and more into science and math. But this return to a secular worldview was not permanent. Although some historians question this, I believe (as reported before) that his short sojourn in Prague in 1911 was crucial for a reengagement with his

Jewishness. Here are some more details about that group of Jewish intellectuals.

First, they were diligently studying social and theological matters around Zionism and mysticism, with special interest in the ideas of the 17th century Jewish philosopher, Baruch Spinoza. On the topic of Zionism and mysticism, there is some debate as to how much this influenced Einstein.[44] It is true that he was not strongly attracted to Zionism until after his move back to Germany in 1914 and his exposure to the endemic anti-Semitism. Nonetheless, the seeds of Zionism were sown in Prague and, as will be seen, came to fruition in the 1920s; especially when he toured the USA with Chaim Weitzman of the World Zionist Organization in 1921, raising funds for the building of the Hebrew University of Jerusalem.

As for any attraction to mysticism, Einstein's scientific worldview precluded flirting with anything supernatural, psychic, or magical. In the 1920s, he referred to the vogue of various forms of spiritualism "as a symptom of confusion and weakness." Even psychoanalysis, which Sigmund Freud asserted was grounded on scientific principles, Einstein had serious misgiving about; indeed, he refused an offer to be psychoanalyzed, saying that he would rather remain the dark.[45] The Prague group's interest in mysticism probably meant they were reading medieval Jewish writings, such as the Zohar and Cabbala literature. In a letter, Einstein once referred the Prague group as "a small medieval-like band of unworldly people."[46] The medieval reference may be related to this literature, and as far as "unworldly" goes, it surely could apply to Einstein himself. I imagine he enjoyed the camaraderie of these marginal folks in an otherwise unpleasant Prague, despite not agreeing about all their reading materials.

If for no other reason than the group's focus on Spinoza, Einstein surely would have been drawn toward them, for he had read the 17th century Jewish philosopher's *Ethics* in the Olympia Academy days as reported by Solovine, and was fascinated by him.[47] Later in life, Einstein made it clear that Spinoza was his favorite philosopher, and often quoted him when discussing theological matters. Over his life, Einstein was influenced by a range of philosophers in both his scientific work and his theological speculations. In the course of this intellectual journey, however, most were put aside, as he developed a more mature and nuanced view on such ideas. But Spinoza's star never dimmed.[48]

By the 1920s, when he was questioned about his religious beliefs, he invariably said he believed in Spinoza's God, "who reveals himself in the harmony of all that exists, but not in a God who concerns himself with the fate and actions of human beings."[49] Spinoza was excommunicated from his Jewish community in Holland for such ideas, which were, in part, seen as pantheistic; that is, the idea that God is identical to the universe. In traditional religion, God transcends (or lies beyond) all that is; therefore, pantheism was deemed as another form of atheism. (In contrast, atheists view pantheism as just another form of theism, or a belief in God.) That Spinoza was therefore an outcast surely was another factor in Einstein's attraction to him.

Accordingly, there were critiques leveled against Einstein by theists accusing him of being a pantheist. In one sense, he was not, for when he said that God "reveals himself," he seemed to imply a transcendent God. Here is how he put it: "I am not an atheist. I do not know if I can define myself as a pantheist. The problem is too vast for our limited minds." [50] Perhaps the simplest and most direct way he expressed his belief comes across in this story. In 1952, in his office in Princeton during a visit by a Jewish student from Yeshiva University, he was asked if he believed in God. Einstein's response began with a waving of his hand and pointing toward the window, which looked out onto a pastoral scene: "All that is not an accident," he said. [51] 'Nuff said?

The 1914 move back to Germany immersed Einstein into the social and intellectual life of Berlin. As seen in Chapter II, the idea of bringing him to Berlin was initiated by Walther Nernst, and he and Planck made the trip to meet Einstein in Zürich. What was not mentioned there was that the physical chemist, Fritz Haber, who was Director of the Institute for Chemistry in Berlin, played a key role by meeting with the Minister of Education with the proposal to create the post for Einstein in Berlin. Haber had met Einstein in 1911 at a scientific conference and they struck up a friendship. Ten years older than Einstein, Haber was born Jewish but converted to Lutheranism at age twenty-four in order to fit into German society and evidently to avoid the obstacles in employment for most Jews. For him this move was also coupled with a strong sense of German patriotism that was common among Jews who felt they were part of the German nation.

Going back to the 19th century, such conversions were rare but not uncommon among German Jews, especially those whose talents were otherwise thwarted by latent anti-Semitism in society. The conversion, however, was not always a ticket to acceptance. As Haber, early in his career, once wrote to a colleague, "It is very difficult for me to get a chair anywhere. ... Jews or baptized Jews are not wanted in the major positions."[52] Nonetheless, as Jews were gradually acculturated into some pockets of German intellectual and social life, they often felt they had more in common with their Christian acquaintances than with Orthodox Jews. This was especially true when there was an influx of Eastern European Jews from the Shtetls (Jewish ghettos) in the 1920s.

KWIK, and importantly, Einstein never even considered the option of conversion, despite his ambivalence with Judaism at different times in his life. The idea, I believe, would have been anathema to his attitude toward authority. He also abhorred many aspects of Germanic culture; the zealous militarism is an obvious one. But also the male rituals of dueling and excessive drinking. [53] Recall that in Aarau, Einstein the teetotaler was reading Kant while his fellow students were consuming beer. In an essay published in 1934 he spoke of a baptized Jew of present and past as a "pathetic creature." [54] Did he have his friend Haber in mind? There is no evidence that Haber ever discussed his conversion with Einstein. Haber, interestingly, despite his conversion, had almost exclusively Jewish friends. [55]

III. Race

As a scientist, Haber went on to receive the Nobel Prize for Chemistry in 1918 for his synthesis of ammonia from nitrogen and hydrogen, a discovery that revolutionized the production of fertilizer, making high yields in agriculture throughout the world possible. The other application was to make explosives.

When the Einsteins moved to Berlin, they struck up a friendship with Haber and his wife, Clara, who had also converted. As Albert and Mileva's marriage unraveled, the Habers were in the middle, negotiating for both sides, and they continued mediating after Mileva moved back to Zürich. Then when the war broke out, despite their friendship, the fundamental differences between Einstein and Haber came to the fore. Einstein viewed the war enterprise as mass madness, and did not sign the patriotic document supporting Germany's invasion of Belgium. As seen in Chapter II, he signed instead a counter-manifesto, and began working with the pacifists. Haber, conversely, signed the official one, along with Planck, Nernst, and ninety other intellectuals.

With patriotic verve, Haber put his full-time effort into the war. Little did he know that as early as August 24, 1914 his friend Einstein wrote in a letter that "the best talent is being forced into this senseless butchery and henchman's service."[56] Haber eventually became the chief war scientist, directing work in explosives and was instrumental in the development of chlorine and other poisonous gases used in the war (think of mustard gas). His work in fertilizers displayed a benevolent side of science, but this chemical warfare exposed a malevolence goal – an endeavour that Haber's wife, Clara, could not live with. In 1915, she shot and killed herself with his pistol. This was the first family tragedy emanating from his work on poisonous gases. Later, as an even more egregious consequence, Haber's Institute went on to develop, as a pesticide, Zyklon B, which in the 1940s was used in the Nazi gas chambers to kill some of Haber's friends and relatives, along with millions of others.

World War I unleashed the latent anti-Semitism in German society when Germany lost the war. It came out of the shadows as

Socialists and Jews were blamed by right-wing Nationalists for Germany's defeat. No longer could many patriotic German Jews believe that anti-Semitism was a mere anomaly that would go away with more assimilation. Those who were never fully comfortable with the contradiction between their otherness and their overt patriotism often turned toward Zionism. [57] Einstein, as seen, was exposed to the Zionist ideal during his Prague sojourn; and so, in Berlin, he again befriended a group of Jewish intellectuals devoted to the cause of Zionism. During and especially after the war, Einstein's identification with Judaism grew as he increasingly referred to himself as being a member of "the tribe."

The growing and explicit racism in Germany was directed increasingly toward Einstein when he became famous following the Royal Society of London's experiment during the solar eclipse of 1919, initiated by Arthur S. Eddington, which proved, as Einstein had predicted, that light from a star is bent by the sun. A Jew getting so much attention and adulation grated on the bigots, some of whom were his fellow scientists. Surely this behavior was most troubling for him, since he expected more objectivity from them. Of course, there were legitimate questions to be raised about the theory of relativity, but many of the attacks were simply racism masquerading as a critique of relativity. The vitriolic condemnations of relativity in 1920 were the start. There were two episodes.

The first began in the summer at an anti-relativity rally in (of all places) the auditorium of the Berlin Philharmonic organised by a right-wing political group. Einstein was accused of plagiarism and propaganda. The former was a red herring, since all scientific work is based on previous work. Real plagiarism would require copying entire sections from other writings, and this Einstein never did. His work was, in part, original in the way he put together previous ideas. The so-called carp about propaganda was bizarre and exposed the blatant anti-Semitism of the whole rally. Supposedly the theory of relativity was being disseminated by mainly newspapers and publications associated with Jews. Whether true or not, the topic was irrelevant to the veracity of the theory. This irrelevant attack showed how desperate the group was to cook up a critique. The real sinister point of this denigration of the theory was to create an etymological dichotomy between supposedly two types of science: German and Jewish. The so-called German science would later morph into what was called Aryan science, as the pseudo-scientific ideology about an Aryan race grew deeper roots, especially after the Nazis took power in 1933.

The wrangle over relativity was more focused in the second episode in the fall of 1920. It came during a scientific society meeting, where a debate was arranged between Einstein and Philipp Lenard. Lenard's name has come up before, since Mileva once attended a lecture by him and Einstein used Lenard's work on the photoelectric effect for his famous paper on light quanta. From the reports of the debate, we know that philosophical issues were at the heart of the theory. [58] However, what is important for this chapter is that the debate later became framed in the combative rhetoric of two conflicting categories of physics: German Physics (Lenard) vs. Jewish Physics (Einstein), as if there is an ethnic (really racist) basis to the way science is done. [59] The sinister side of this was made manifest in subsequent years when Lenard, along with Johannes Stark, joined the Nazi party and they attacked Einstein as expounding a Jewish Physics. Lenard put forth a warning about an "alien spirit ... which appears everywhere as a dark force and which leaves its mark so clearly on everything that belongs to the 'theory of relativity'." [60] For such rhetorical gibberish, Hitler would give Lenard the highfalutin post of Chief of Aryan Physics. [61]

As the drama over relativity was playing out in the 1920s, Einstein took up another cause – the plight of Eastern European Jews. Since the late-19th century, waves of Jews fled pogroms in Russia and passed through Poland to arrive in Germany. During World War I they were recruited to work in factories for the war effort under appalling conditions. After the War some continued onward to the USA or to Palestine. Those who stayed were targets of anti-Semitism, for they were poor, usually working as peddlers, and seen as "parasites" in German society, or worse as "vermin" – a term portending the racism of the Nazis. Eastern Jews were also often avoided by the "western" German Jews, who had integrated into German bourgeois culture and had a hard time identifying with these ghetto-living, Yiddish-speaking, skullcap-wearing co-religious folk. German authorities sometimes went so far as to put hundreds of Eastern Jews into internment camps, a move that foreshowed the concentration camps of the Nazis.

Einstein, who invariably took the side of the underdog, could not quietly watch this blatant act of discrimination of a minority, especially members of his tribe. Using his new status as a celebrity with a public voice, he wrote in support of the Eastern European Jews, and spoke out repeatedly against the rise of anti-Semitism. He challenged the German government to stop this discrimination by reminding them that during the War the Germans were accused by the enemy of acting like barbarians (which they really did in Belgium), and so in their treatment of Jews they were reinforcing their own stereotype. He weaved an image of these Jewish peddlers as not hordes of beggars but rather as "a wealth of the finest human talents and productive energy" (which, in the long run, turned out to be true, at least for those who were not murdered by the Nazis). Einstein then put this idea into practice by helping to organize special lectures for Eastern Jewish students at the University, who otherwise were not admitted. [62]

This involvement in the plight of Eastern Jews was another element in his reengagement with his Jewish identity at this time. Coupled to this was his increasing involvement in one aspect of the Zionist movement – something during the Prague sojourn he only viewed from a distance – namely, the quest to create a Hebrew University in Jerusalem. In fact, he became downright zealous about it. As he wrote, "Many talented Jews are lost to culture because the way to learning is barred to them. It will be one of the foremost aims of the university in Jerusalem to alleviate this misery." [63] The only barrier would be a lack of talent, not ethnic origins. KWIK, he probably had Eastern European Jews in mind when he wrote this.

The idea for such an institution went back to the late-19th century, and by the 1913 meeting of the World Zionist Congress it was decided that a University whose language of instruction would be Hebrew should be built. In 1918 the organization obtained permission from the British Commonwealth to lay a cornerstone on Mount Scopus in Jerusalem, and they did so in the summer, on July 24th. Einstein thus envisaged a place where Jews from anywhere in the world, initially from Eastern Europe, could study freely, with admittance based on merit alone.

In 1921, Einstein made his first trip to the USA on a tour in support of the Zionist movement, with the specific goal of raising funds for the Hebrew University. Chaim Weizmann, chemist and president of the World Zionist Organization, prodded him to do so. Weizmann wanted Einstein to join him on this tour, since his celebrity status surely would help to draw crowds and hopefully increase donations. Einstein was not naïve about all this. He wrote: "I am not eager to go to America but am doing it solely in the interest of the Zionists, who must beg for dollars to build educational institutions in Jerusalem, and for whom I act as high priest and decoy ... I am really doing whatever I can for the brothers of my race who are treated so badly everywhere." [64]

The 1921 tour was from April 2nd to May 30th, and he was often greeted at stops with much fanfare and incessant questions from reporters. Albert Einstein, now the eponym *Einstein*, was learning to adapt to the role of celebrity and the accompanying lack of any privacy. All was not just fundraising, however; he lectured at several universities on relativity. At Princeton University he gave a series of lectures that were published as the book *The Meaning of Relativity*, a work that is still in print. [65] Of course, he did not know that starting in 1933, he would spend the rest of his life in the town of Princeton, New Jersey, not far from the University. By the end of the 1921 trip Einstein was deeply devoted to the cause of creating a Hebrew University in Jerusalem.

As noted in Chapter **II**, in 1920, the tension and distress over the attacks on Einstein were, at least partially, responsible for

Elsa's ailments at this time, and led to the couple's serious thoughts of leaving Germany. Yet they stayed. They stayed despite, for example, in June 1921, after Einstein returned from this USA tour, there appeared an article in a Nationalist newspaper calling for his outright murder. Events like this spurred him further into the Zionist fold. Later in the month, he delivered a speech to a Zionist meeting in Berlin, in which he concluded:

> If we could succeed in establishing a center for the Jewish people in Palestine [then] we will have again an intellectual center and the feeling of isolation will leave us, despite the fact that most of us are scattered in all countries. This is the great liberating effect that I expect from the rebuilding of Palestine. [66]

It was followed by much applause.

In the spring of 1922 Einstein made a trip to Paris for a series of talks and seminars on relativity. When first approached by French scientists he rejected the offer, due to the lingering animosity between the two countries since the War. Johannes Stark, for example, saw it as a capitulation with the lingering enemy.[67] But Einstein was friends with Walther Rathenau, Germany's foreign minister, who saw the trip as a positive effort to mend the rift. Rathenau was, like Einstein, a Jew who was "internationally minded," as Einstein once put it. But Rathenau was also very much unlike Einstein, for (again quoting him) Rathenau "was very much in love with Prussianism, ... and its military forms." [68] Einstein was probably attracted to Rathenau because of his intelligence and especially his wit and the subtle ways he could make disparaging remarks about social mores. Einstein, in the end, agreed to go to Paris. A French delegation met his train at the Belgium border to escort him to Paris; they were concerned about security, for there were threats made by some French ultra-nationalists opposed to a

German visitor. As well, some French scientists boycotted a reception.

The trip, however, went off without any serious incident, and after he returned there was a German-French friendship rally at the Reichstag (the German Parliament) in June, where Einstein was met with much applause. But the hoopla was short-lived. Two weeks later Rathenau's car was riddled with submachine-gun bullets and a hand grenade was thrown in to make sure he died. This cold-blooded murder of his friend both angered and rattled Einstein. He contemplated being the next Jew assassinated by right-wing thugs, as the death threats resumed. In a letter to Solovine he said that he cancelled all his lectures and was officially absent from his office, whereas he was "actually always here all the time."[69]

He and Elsa again seriously considered permanently leaving Germany, but instead went on a trip. In October, they commenced a five-month tour to the Far East. Being a world-famous celebrity, Einstein was taking advantage of an abundance of travel invitations and the opportunity to leave the turmoil of Europe for periods of peace elsewhere. On their return to Europe they passed through the Suez Canal and made a scheduled stop in Palestine for a two-week tour. They visited most of the landmarks of the region (such as the remaining Temple Wall in Jerusalem) but he was particularly interested in the kibbutz movement – the Jewish collective farms based on the communist ideal, which feed right into his flirtation with various segments of socialism.

He was emotionally affected more directly and deeply in this leg of their voyage than anywhere else, mainly because at this stage of his life, with the exposure to anti-Semitism in Germany, he was probing deeper into his Jewish roots. Yet it was a completely different Jewish world he met in Palestine, a far cry from the lifestyle of both the European professionals and the Eastern peddlers. As he wrote in a letter: "The brothers of our race in Palestine charmed me as farmers, as workers, and as citizens."[70] At one point in the tour, among the adulations of a large group of school children and teachers, he was quoted as saying it was "the greatest day" of his life. [71]

As part of his visit to Palestine, he was to give a lecture to the nascent Hebrew University. He did so on February 7, 1923 in a temporary building (in a hall of the British police academy on Mount Scopus in Jerusalem), near where the cornerstone was laid in 1918. He began his speech with an arduous introduction in Hebrew, so that the first words spoken in the nascent university, indeed, would be in that ancient language. He continued in French, and went on for ninety minutes to deliver essentially the first lecture at what eventually became the Hebrew University of Jerusalem. The topic was an outline of the theory of relativity.[72] Two years later the University was officially and formally opened in an historic ceremony on April 1, 1925. Einstein was not present, but not unexpectedly, when the first Board of Trustees for the University was announced, he was made a member. [73]

As seen, Einstein waffled during his early life with his Jewish roots.[74] It was in the early 1920s, perhaps as a result of the Palestine visit, that he settled into a comfortable place with his Judaism. He identified as a Jew culturally and ethnically, but from a non-religious (especially non-Orthodox) viewpoint. On the Palestine tour, for example, at what remained of the ancient Temple Wall destroyed by the Romans in 70CE, where he saw Orthodox Jews praying aloud, he wrote in his dairy that these "dull-witted clansmen of our tribe" were a "pathetic sight of men with a past but without a future."[75] This was, of course, in contrast to the Jews on the kibbutz.

In addition to his embrace of Spinoza's theology, he eventually developed his own socio-ethical creed, succinctly put in this sentence from the 1930s:

> The pursuit of knowledge for its own sake, an almost fanatical love of justice, and the desire for personal independence – these are the features of the Jewish

tradition that make me thank my lucky stars that I belong to it. [76]

Einstein had made peace with his Jewishness. But he was yet to reconcile his internationalism and pacifism with the militarism and the belligerent anti-Semitism in the country in which he lived.

Upon returning to Germany in 1923, the intimidations resumed. As mentioned before, during an academic sojourn in Leiden, he stayed for six weeks because of death threats in Berlin. [77] For most of the 1920s, the trips outside of Germany assuaged the omnipresent anti-Semitism he was forced to endure. In addition to the Far East tour, there were lectures in South America, the United Kingdom, and the USA. In the USA they spent three winters in California from 1930 to 1933, the final one during January 1933 when Hitler and his henchmen came to power. As a result, the Einsteins finally made the inevitable decision to live in the USA. How many death threats can a person tolerate? They never returned to Germany.

Upon assuming power over the German government, the Nazis began enforcing their racist laws. By the spring of 1933, there was a very significant decrease in professors at German universities. The numbers are mind-numbing: 10% of all professors lost their positions because they had "Jewish blood"; this was 20% of mathematicians and 26% of physicists. It was the start of probably the greatest brain-drain in modern history. John von Neumann, a famous Hungarian-American mathematician who was studying in Germany at the time, wrote in the summer of 1933 that this "German madness…will ruin German science for a generation – at least …" It came true: by the spring of 1936 more than 1600 scholars (1/3 scientists) left German institutions, going to the USA, UK, Canada, and elsewhere. The long-term effect can be measured, in part, by comparing Nobel Prizes in science before and after World War II. The Prize started in 1901, and by the War, the score was: Germany 35, USA 15. After the War, through 1959, it was: Germany 8, USA 42. Many of the 42 were, without doubt, transplanted refugees.[78] Calling it madness is not strong enough. Einstein called the rise of Hitler "mass psychosis." [79] Kicking out the smart people and giving the country over to hooligans is sheer mass stupidity too.

In October, 1933 the Einsteins moved to Princeton, New Jersey, where he took up his post at the new Institute for Advanced Study. Otherwise, if they had remained almost anywhere in Europe, Einstein (and probably Elsa too) most surely would have been assassinated or murdered sometime during the Third Reich's reign of terror. It is a horrendous fact that the attempted extermination of European Jews by Hitler was not confined to Germany itself, but extended into the conquered states.[80]

Einstein, accordingly, revised his pacifist position. The German threat was too great, and a military force opposed to Nazi aggression was necessary. His pacifist friends, having lost such a powerful voice, were aghast and angry, but he made it clear that this was an exception – for no principle is absolute (in both science and ethics). As he later wrote, this "exception" was "necessary," especially with such a "hostile power" that endeavoured the mass extermination of his "own group." [81]

Haber's final story is tragic. At the end of World War I, Haber was a war hero, with a long and distinguished record of service. That record was entirely erased from German history books when Hitler came to power in 1933, as if Haber never even existed. All that exuberant patriotism, the conversion to Christianity, the assimilation into German society, the accolades – it all came to naught in the warped minds of the Nazis. They did not kill Haber directly, as they would have eventually. Instead, he fled Germany but died of a heart attack in Basel, Switzerland a year after Hitler took power. His eventual disillusionment with Germany was expressed, rather mildly, in a letter penned shortly before his death: "Lucky the person who did not grow up in the German world..." [82] His son, Hermann, found refuge in the USA, but committed suicide, as his mother had, in 1947. Shortly thereafter, Hermann's oldest daughter did the same.

Searching for a glimmer of light in this otherwise wretched story? After Haber's death in 1934, Planck attempted to keep Haber's memory alive by courageously holding a memorial to him in Berlin that 500 people attended. It was a rare act of defiance by a scientific community that otherwise easily buckled under the bullying of the Nazis.

At this juncture in my story it is at once enlightening and depressing to bring this topic up to the present with the following facts. If you search the Internet for information on Einstein, you will, more often than you may wish, come upon a website that initially looks as if it is a scholarly discussion of a topic related to him, but find, on deeper reading, that you have been lured into an anti-Semitic diatribe castigating Einstein and Jews. It is disconcerting how frequently this happens, as one innocently searches the Web.

I will not give credence to this evil by citing any such site, except to mention one related website, for which there is a significant audience. It is put out by a group of right-wing, ultra-conservative Americans, who believe that Wikipedia is a liberal-biased source of left-wing information. Called Conservapedia, it is an ideologically based encyclopaedia that openly attacks Darwin and evolution (no surprise here), and adds Einstein and relativity to its list of iniquities. [83] The articles related to the latter pair echo the anti-Semitic Aryan attacks of the Nazi era, denying any originality to all of Einstein's work, and challenging the empirical basis of the theory – all of which contradicts volumes of scholarship and reams of experimental data. The articles are rubbish.

Decades of 20th century scholarship on Einstein have shown his indebtedness to former and contemporary scientists, while revealing the originality and genius of what he did. This was realized by the second decade of the last century. Consider the following document. In July 1913, Planck, Nernst, and two other physicists signed a letter of recommendation to the Prussian Academy of Science supporting Einstein for the job in Berlin. In it they spoke of his "worldwide reputation" around his relativity theory that presented a new "conception of time." He also contributed to the quantum theory, the kinetic theory of matter, atomism, and thermodynamics. They pointed out that he was involved in almost every important physics problem of his day, revealing not only his depth, but his breadth, as well. [84] If Einstein had been stealing ideas from others around him, at some point these fellow scientists would have known. Nonetheless, the myth of plagiarism persists. The unsettling nature of all this prompts me to quote Philipp Lenard (of all people!) in an entirely different context, and speak of an "alien spirit, which appears everywhere as a dark force and which leaves its mark" – a stain that shamefully just will not go away.

Einstein's experience in Germany shaped his outlook on matters of race throughout his life. When moving to the USA he was confronted with another form of bigotry and intolerance in the way African-Americans were treated. He became visibly involved in the civil right movement, using his celebrity status in support of their cause.

Yet, even prior to the move to America, in 1931 he joined an international campaign in support of the "Scottsboro boys," nine African-Americans falsely accused of the rape of two white girls and sentenced to death. (It took nineteen years before they were finally set free.) As well, in 1932, during his last trip to California, Einstein attended a memorial service in a Black church. The service was honoring a Jewish philanthropist who was active in supporting education for African-Americans. Einstein delivered an address calling for racial tolerance and world peace. [85]

Later that year he received a letter from the noted African-American, W.E.B. Du Bois, author, activist, and co-founder of the NAACP (National Association for the Advancement of Colored People). A prolific scholar, Du Bois became aware of Einstein's viewpoint on race and so he asked for a word from the famous scientist. Einstein replied with a brief statement on the tragic and evil nature of the minority status of the "American Negroes," which was then published in *The Crisis,* the monthly magazine edited by Du Bois.

Einstein went on to befriend openly other black Americans, such as the opera diva, Marian Anderson, the popular singer and actor, Paul Robeson, and others. He met Anderson in April, 1937 when she gave a concert at a Theater in Princeton, receiving rave reviews; nonetheless, she was refused lodging at the local Inn. Einstein thus invited her to stay with him on Mercer Street, and she did so every time she came to town. They remained life-long friends. This was also true for Robeson, with whom he co-chaired the "American Crusade to End Lynching," an organization that spoke out against the nefarious vigilante hangings of black Americans, mainly in the South. Indeed, this despicable act was targeting returning veterans from the War, who otherwise should have been revered as war heroes. Einstein was appalled and spoke out. For this effort, he, Robeson, and the anti-lynching organization was investigated by the FBI as a probable Communist front.

Einstein's general position on racism may be found in an essay published in January, 1946, "The Negro Question." [86] He began by pointing to his immigrant status in the USA, suggesting that it could be used against him for making critical comments about his country of choice. Yet, he believed that this status, in reality, gave him added insight into seeing things that others might not, such as taking prejudicial behavior for granted. Therefore, despite the "democratic trait among the people" of America, where "everyone feels assured of his worth as an individual," Einstein noted "a somber point in the social outlook of Americans": namely, the injustices against non-whites, "particularly towards Negroes, a situation," he wrote, that "pains me." Speaking to the whites of his adopted country, he candidly stated: "Your ancestors dragged these black people from their homes by force" and "ruthlessly suppressed and exploited" them, such that they were "degraded into slavery." The legacy of this still remained in the "deeply entrenched evil" of racism in America. As always, Einstein did not mince words.

After he moved to the USA Einstein was reluctant to accept honorary degrees. He was tired of the tedium of the pomp and pageantry of it all. But in 1946 he made a rare exception when he accepted an invitation from Lincoln University, an all-black college, the first such institution in the country, which was about sixty miles from Princeton in the south-east corner of Pennsylvania. He wanted to show by his action, not just his pen, his support for the civil rights movement.

He was invited by the President, Dr. Horace Mann Bond, grandson of slaves and father of the later civil rights leader Julian Bond. (Julian was six at the time of the visit, when Einstein gave him this advice, which he says he followed: "Don't remember anything that is already written down.") The event took place in one day, on Friday, May 3rd, 1946. It was held outdoors since there was no hall large enough to hold all who came. The student newspaper wrote that Einstein was as "a very simple man," who appeared as "a biblical character ... with an expression of questioning wonder upon his face" as President Bond read the citation of the honorary degree. Upon receiving the degree, Einstein was reported to have said that segregation "is a disease of white people," and he added: "I do not intend to be quiet about it." Much of this echoed his essay from January of the same year.

Sadly, there is no copy of his speech to the university, since the event was barely mentioned in the white press (except for a minor note in the *New York Times*). This is in contrast to the habitual behavior of the American press, which strove to record seemingly every word coming from the white-haired sage on Mercer Street – as if he were a font of wisdom on everything. So what happened when he spoke out about the endemic American racism? There was a curious and deafening silence. Thus, what is known about Einstein's visit to Lincoln University, as quoted above, was gleaned from the student newspaper and the black press, where it was widely reported. Later in the day he gave a lecture on relativity to the students, before for he departed the campus. [87]

Later, in the fall of 1946 there was a large multi-faith, inter-racial rally, including both black and white veterans of war, at the Lincoln Memorial in Washington, D.C. Robeson was there and sang. Einstein was scheduled to speak, but illness prevented his appearance. Various reports indicate that it drew thousands of protesters. Such a large-scale event portended the civil rights rallies of the 1960s. This and the Lincoln University visit, demonstrate the role race played in the socio-political world in which Einstein lived during his years in America.

The next major event affecting Einstein's life was the creation of the state of Israel in 1948, coming after the almost impossible comprehension of the horror of the Holocaust. The long struggle for the realization of Zionism was made real. Einstein was elated, but ambivalent. When he was asked to be the second president of Israel (a ceremonial job), he declined.[88]

As a teenager he had fled the militaristic nationalism of the German state and renounced his citizenship. He therefore had difficulty living in such an environment when he returned in 1914 and lived through the excessive patriotism of his colleagues and some friends during and after World War I. As seen, that experience tainted his flirtation with Zionism. He wrote in 1929 that his idea of Jewish nationalism was "a nationalism whose aim is not power but dignity and health. If we did not have to live among intolerant, narrow-minded, and violent people, I should be the first to throw over all nationalism in favor of universal humanity." [89]

During his visit to Palestine in 1923, he saw the best element of Jewish society in the working class, what is called Labor Zionism. As he put it in 1932: "This working class alone is the only force which is capable to create healthy relations to the Arab people, which is the most important task of Zionism." [90] His viewpoint on the Palestine problem, then and throughout his life, was grounded on a nuanced attempt at balancing the rights of both Jews and Arabs to the same relatively-small parcel of land. He put it strongly in 1938: "I should much rather see a reasonable agreement with the Arabs on the basis of living together in peace than the creation of a Jewish state..." [91]

He envisioned Israel as a cultural and spiritual center of Judaism, not a political entity. Nonetheless, this idyllic arrangement was not to be. Following World War II the push for a Jewish state was amplified, and he conceded in 1945: "I dislike nationalism very much – even Jewish nationalism. But our own national solidarity is forced upon us by a hostile world." [92] Einstein could be stubborn on many things, but he also was not dogmatic, and was willing to modify his position if it was deemed reasonable by him. Just as his strong commitment to pacifism during World War I was severely revised with the rise of Nazism, he felt that the nationalism around the state of Israel was an exception to the rule.

This fact provides a segue into the next chapter, for this non-dogmatic, quasi-flexibility was also applied to his science. He occasionally conceded that if future experiments would falsify his theory of relativity, then the whole edifice should rightly come tumbling down. In short, he accepted the possibility (remote, he thought, I assume) that his entire life's work in science could be all for naught.

IV. Chutzpah

Chutzpah: a Yiddish word; the *ch* is not pronounced like *church* but like the Scottish *loch*; rhymes with "foot spa." Untranslatable, but its meaning is close to these English words: audacity, gall, effrontery, or guts (Rosten, 1970). Having moved into the popular vernacular, chutzpah has also taken on the connotation of: courage, resolve, or ardor.

My earliest recollection with something that I would call "science" was in my preschool-years (age five?) when I was told by an adult that the earth is not flat – the way it appears to us – but it is really round. On learning this completely unexpected fact, I remember pondering for some time how this could be true. My thought process went something like the following, which I will illustrate with diagrams, although I do not remember actually making little drawings. I remember these images being only in my mind.

At first, I thought the idea of a round earth should look something like Figure **1**, with me positioned at **A**, on the *outer* surface of a ball or sphere. The problem, in my mind, was that if I moved too far away from this top position, I would fall off. So how could the world be round? Something was wrong with my mental picture. After some time, I came up with a solution to this problem. The solution is hard to picture, but essentially in my mind's-eye I remember having the realization that the roundness of the world went the other way, as if I were living *within* a big bowl, as at position **B** in Figure **1**. On this model, I can move anywhere without falling off the earth. So, I gladly reasoned, this must be the way the earth is shaped, although much bigger than my diagram.

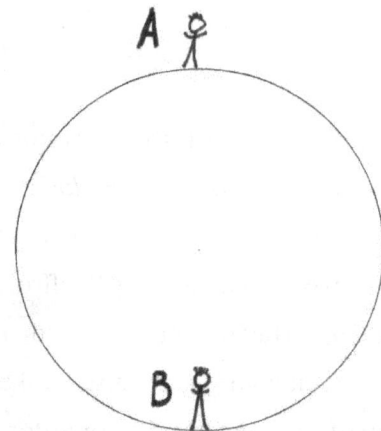

Figure 1
**My conceptual images, at about the age of five,
of how the world could be round.**

On this inside-the-bowl model, if I traveled far enough, the world would curve up and the earth would meet the sky. This at once explained the "roundness" of the world, and the visual dome of the sky, since they both were part of the same mental picture. I surely did not appreciate, as I now do, how clever this idea was. (In mathematical terminology, in going from **A** to **B**, I was making a mental switch from a positive to a negative curvature of space.) For me, I simply believed that I had just figured out how a round world can exist and how it is connected to the sky.

In addition, what I obviously did not know was this: at about the same time when I was having the first scientific thought of my life in Pittsburgh, Pennsylvania – that Albert Einstein, 340 or so miles away in Princeton, New Jersey, was contemplating the structure of the universe in similar terms of either a positive or negative curvature or maybe just a flat space.[93] How strange (or fascinating?) is the realization that I was (sort of) thinking along similar lines as Einstein for the structure of the world. Gives me goose bumps.

Einstein created modern cosmology in 1917. It was a theory of the structure of the entire universe. One of the questions that grew out of this work, as Einstein and other physicists and cosmologists explored the model over the years, was the shape of the space of the universe, for which there were three possibilities – positive, negative, or flat. Einstein was still thinking about this in the late-1940s – when I was absorbed in my first thought experiment – and it still was not settled when he died on April 18, 1955. Today the consensus is that the space is flat, although there may be some positive curvature too.

Einstein's paper of 1917 was a bold act. The universe since Isaac Newton in the late-17th century was conceptualized as an open expanse of space filled with stars going off infinitely in all directions of 3-dimensional space. That image was understood throughout the 18th and into the 20th century as an undisputed fact, supported by increasing information from astronomy, as bigger and better telescopes continued finding new things in the heavens. There were, however, some questions about how much of the infinite universe we can actually know. For example, some serious astronomers believed that we were trapped in our Milky Way, and that it was *all* we can ever know; the infinite space beyond, although true, was the realm poets and mystics, not scientists. Despite this possible observational constraint, most scientists continued to speak freely of an infinite universe.

The initial acknowledgment and belief in this infinite image of the universe by astronomers and everyone else required a major revolution in thinking that took place about 350 years ago. The common sense idea of the earth at the center of the universe, with the stars being attached to a finite sphere that rotates around us, was a picture of our place in the universe that all cultures on our entire planet conceived of whenever they contemplated these fundamental and profound questions: What is the universe and where are we? Today we still create an illusion of how the heavens appear to us using this model, as when we sit in a planetarium under a real dome with lights projected onto its surface imitating stars and other celestial objects.

It took a litany of brilliant thinkers from Nicolaus Copernicus, through Galileo Galilei, Johannes Kepler, Newton and many, many more to convince Europeans that a counter-model, with the stars scattered in an infinite space, was really how the universe was. It was this model that Einstein and all other children were taught in school in the late-19th century, an idea that was reinforced as being unchangeable after centuries of evidence. Yet change it, he did. Einstein's 1917 model, like my imaginary round earth, took the flat space of Newton and bent it into a 4-dimentional (4-D) sphere. As a way of picturing this model, consider the case of our 3-D world viewed only as 2-D shadows cast on a wall. By analogy, Figure **2** would be the 3-D shadow-world of Einstein's 4-D model, with we humans (see the flat guy with the curly hair embedded in the space) as 2-D beings. Such a curved world would be finite, unlike Newton's infinite model, and any voyage in a straight line, eventually, would come back to the starting point, as the arrows in the Figure suggest. (This is analogous to someone traveling along the arc of a great circle on the earth.) Here was a radically different model for modern cosmologists to ponder, since Einstein had put forth a new universe – unlike none other before. [94]

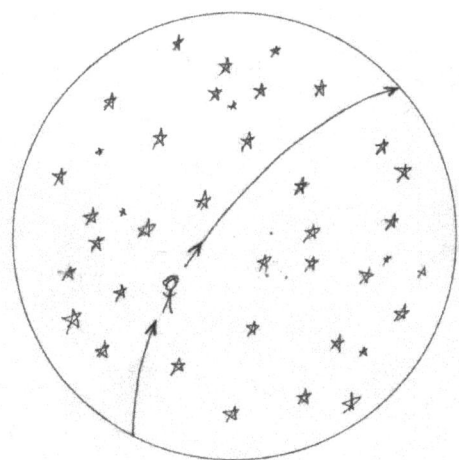

Figure 2
The 3-D analogy of Einstein's 4-D model of the universe as curved and finite.

In many ways, the 1917 cosmic model was the culmination of his theory of relativity – a theoretical framework that began a dozen years earlier in his famous publications of 1905, or even a decade before that at Aarau, with the thought experiment (mentioned in Chapter I) about riding a beam of light. What was it, and how did it lead to relativity?

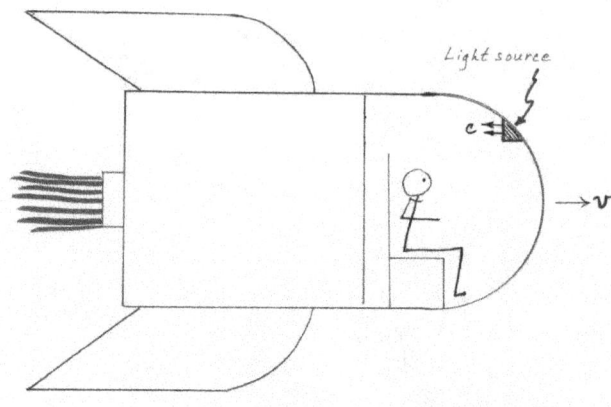

Figure 3
My reconstruction of Einstein's thought experiment about traveling at the speed of light.

Einstein never clearly explained how that famous thought experiment worked, since his short description in his autobiography is cryptic. Here, however, is (I hope) my trustworthy reconstruction, using Figure **3** showing a spacecraft that can reach the speed of light. [95] At any speed (**v**) less than light-speed, the cockpit of the spacecraft would be lit by the light source, sending out light at speed **c**. But if the spacecraft reached light-speed (**v=c**), the light leaving the source could not fill the cockpit, since the craft itself is constantly catching up with the light beam. The traveler would then experience a total darkness in the cockpit. When that happened, the traveler would know he was travelling exactly at the speed of light. And therein was a problem. What problem, you ask? To explain, we need to go back (and very briefly), at least, to Galileo in the 17th century.

For the 21st century thinker, the idea of the relativity of motion is easily comprehended, since the experience is common in most modes of travel today. Unlike 17th century bumpy rides on land

and bouncy rides at sea, the smooth sailing in our airplanes, boats, and cars gives us all the feeling of being at rest with the world passing us by. Cover the windows and it is impossible to know if the vehicle you are in is at rest or moving at a constant speed. (If you are accelerating, by speeding up or slowing down or turning a corner, that is another situation we will analyze later.) Einstein liked to express the idea of the relativity of motion in the terminology of impossibilities: If you are moving at a constant speed, and cannot look outside to see if you are moving or not, there is no experiment you can do to prove you are really moving or at rest. He called this the Principle of Relativity. Whatever you do in a vehicle moving at constant speed, it is impossible to detect if you are at rest or you are moving, since everything behaves as if you are at rest. If you drop an object while moving, for example, it falls straight down, just as it does if you are at rest.

(BTW: It is a trivial fact that this realization is not unique to Einstein. Not only because the idea of the relativity of motion goes back to Galileo, but because around the early 1900s there were others contemplating its meaning for physics. Such a fact provides fodder for the anti-Einstein factions on the Internet and elsewhere who question his singular contribution. However, though there are some minor reservations about certain details comparing how Einstein used the relativity idea compared with other physicists around the same time, the development of what he put forth in his magisterial paper of 1905 far surpassed their work and extended the idea into the experimental predictions as put forth and ultimately confirmed. Moreover, that there was a wider context to his work – that it was part of a much larger scientific, institutional, and sociological domains with many others working in related fields – is a reality still being explored by historians. [96] Like Shakespeare, Einstein is an almost endless source of deeper and further investigation and interpretation.)

Returning to Einstein and the relativity of motion: we go back to Figure **3** and the thought experiment. For if, at the speed of light, the cockpit becomes dark, then this would be a way of knowing that you are really moving and exactly how fast. This would contradict the Principle of Relativity. So, how can the Principle hold true? After a decade of pondering this thought of travelling at the speed of light, ever since the age of sixteen, Einstein came to the realization that he could keep the Principle by modifying the behavior of light. If the light beam was emitted from its source at the same speed no matter how fast the spacecraft moved, then the cockpit would remain bright and the contradiction would be removed. A simple way of expressing this idea is to say that light always moves at the same speed, which is independent of the motion of the source of light. This unique property of light combined with the Principle of Relativity were the two cornerstones of the first stage, in 1905, of the theory of relativity. He used the format of Euclid taken from the "holy book"; namely, to start with axioms as postulates and then deduce results following the rules of logic. The two cornerstones were the axioms. (BTW: Importantly, this dual combination was also unique in the history of science, which is another reason why I give no credence to the allegation of plagiarism directed against Einstein.) From these two postulates he deduced several strange ways that things in the world would behave if we traveled very, very fast. Such as, time itself would tick off more slowly the faster we moved. Or, the weight of things would increase just by moving. And, most famously, matter (**m**) could be converted into an enormous quantity of energy (**E**), if the matter is totally annihilated ($E = mc^2$). Since the speed of light (**c**) is a very large number, then c^2 multiplied by even a small amount of **m** is still a mammoth number. It was enough to grant anyone fame.

But this first stage of relativity was not all that Einstein did in that year. Here is the list of his accomplishments, in order of publication (only the last two are on relativity):

- A paper applying the quantum concept to light. This paper was a groundbreaking contribution to what came to be the quantum theory of the atom. (This was partially based on the experimental work of Lenard.)

- His PhD dissertation, which he dedicated to Grossmann. In it he presented evidence supporting the existence of atoms, an idea that was *not* commonly held at the time. It was the most cited of all his publications for a decade or more.

- Another paper supporting the reality of molecules or atoms from what was called Brownian motion (that is, the random motion of tiny particles suspended in a liquid or gas).

- His first paper presenting the theory of relativity; the only person he thanked for assistance was Besso.

- A second paper on relativity, this one containing the equation $E = mc^2$.

All this was done while he held a six-day-a-week job at the Patent office in Bern. From what we know from Chapter II, we may view this extraordinary work as the culmination and expression of the real love of his life – namely, physics.

Happily, we know what Einstein thought about all this work in 1905, since he talked about it in a letter to Habicht, of the Olympia Academy. He called the light quantum paper "very revolutionary," although I wonder if he realized how revolutionary it actually was. He believed that in his PhD thesis he measured "the true size of atoms" – a bold statement, since many scientists still did not believe that atoms even existed. This was further supported by his paper on Brownian motion. The first relativity paper he called "only a rough draft" that presents "a modification of the theory of space and time" (indeed!), and the second one was an idea that had crossed his mind.[97] How fascinating. It shows how confident he was in his achievement. In fact, in a draft of a PhD thesis that was rejected, he found a problem in Boltzmann's statistical (gas) theory and castigated him for it.[98] Clearly, for a student, this took guts.

Chutzpah? Yes, it seems. Indeed, let's peruse his work over the next twenty years, as he both developed further his relativity theory and continued to make important contributions (along with other physicists) to quantum physics.

The "revolutionary" nature of his quantum contribution, of which Einstein himself spoke in 1905, was only recently appreciated by historians of science. [99] The first idea of a "quantum of energy" was introduced by Planck in a famous paper of 1900, and for that reason he was considered the founder of quantum theory. But the late-20th century science historian, Thomas Kuhn, eventually convinced the rest of us that Planck's concept of the quanta was not really discontinuous. Rather the concept of a fixed, quantized energy element was in Einstein's papers between 1905 and 1907 and this constituted the real beginning of quantum physics. [100] Furthermore, the corresponding idea of a quantum of light (later called a photon), which he put forth in 1905, was rejected by the physics community well into the 1920s. For example, in the 1913 letter from Planck, Nernst, and others recommending Einstein's membership in the Prussian Academy, from which I quoted in a different context in Chapter **III** – there is, after the list of his accomplishments, this further statement: "That he might sometimes have overshot the target in his speculations, as for example in his light quantum hypothesis, should not be counted against him. Because without taking a risk from time to time it is impossible…to introduce real innovations." [101] Even Niels Bohr, the father of the model of the modern atom, did not believe in photons until the mid-1920s when the experimental evidence could not be ignored. The word photon, therefore, was not coined until 1926; by then, the entity was real, and needed to have a name. Nonetheless, throughout this time, from his 1905 idea of quantized light, Einstein persistently stuck to his certainty in the reality of photons, despite myriad doubters all around him. Chutzpah? Yes, certainly.

The other radical idea of modern science was, of course, relativity. The first stage of relativity (1905), as seen, was limited to objects moving at constant speeds, and became known as special relativity. Starting in 1907 Einstein endeavoured to extend the theory to all motions, such as acceleration, and it became known as general relativity. A brief explanation of what he was thinking about begins as follows.

Newton's idea of gravity was based on the apparent fact that all bodies of matter are instantaneously attracted to each other, such as an apple and the earth, or the earth and the moon, or the sun and the earth. He initially arrived at this idea through a thought experiment about falling objects.

This is the famous story that is often told in terms of Newton sitting under an apple tree in his mother's orchard and seeing an apple fall to the ground. Also seeing the moon (if, say, this were a late afternoon sky), he assumes that the gravity that is pulling the apple to the ground extends all the way to the moon, and this was the great leap of his imagination – or so the story goes. But, as such, this tale makes little sense. What, for example does Newton do next: wait for the moon to crash to the earth? Well, he could wait forever, for the moon does not fall at all.

Nonetheless, this possibly apocryphal story may still be salvaged if we make some modification to it using his own diagram. Figure **4** is my copy of Newton's diagram. It shows his conception of what is the first idea of an artificial satellite.

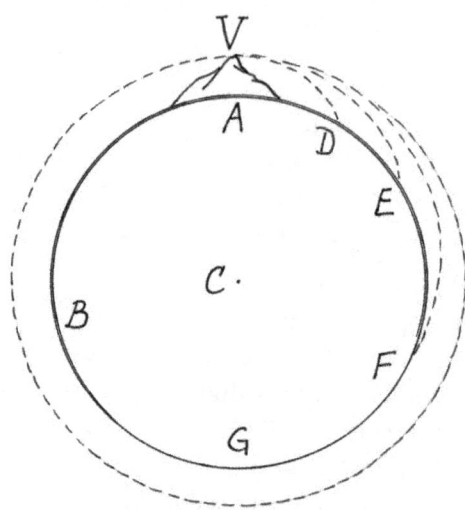

Figure 4
Newton's drawing of his thought experiment of a falling projectile going into orbit around the earth.

It starts with an object launched with a given speed from a mountain top (**V**) at the north pole (**A**) and falling along a projectile path terminating at **D**. There are then a series of launches of horizontal projectiles with increasing speeds that terminate at points **E, F, G,** and **B**. The extreme case is an initial speed sufficient to send the apple all around the earth, such that it returns to the place of launching (**V**); in this case the apple would go into orbit. Gravity is still pulling the apple to the earth, but it never reaches the ground, so that it falls forever (of course, only if there is no resistance of a medium). This thought experiment by Newton was a monumental achievement.

From this conception, it was a short leap to the idea that the moon was such an orbiting object, and hence was being attracted to the earth by gravity. So gravity did extend to the moon; it was not confined to the local shell near the earth. Furthermore, since the earth, along with the planets, moved around the sun, perhaps this same gravitational attraction acted among all these objects too, as if the planets of the solar system were all satellites of the sun. Later as he developed his theory mathematically, he made the final leap to what is called universal gravity, filling all space out to the furthest stars.

As a scientist, however, he found this behaviour strange, for how could an apple know where the earth was and vice versa? Or the sun and planets? Or the stars, being so far apart? This became known as the problem of action-at-a-distance. These bodies, not only close by, but at very far distances, seemed to be instantaneously communicating with each other. How could this be? It was mysterious, if not mystical. Newton never came up with a satisfying explanation for the cause of gravity. In the end, he believed that gravity was caused by God.

This mysterious attraction bothered Einstein too, who once called gravity "spooky." [102] But in 1907, he came up with a way of getting around this problem, and the result eventually was his general theory of relativity. Before we look at the genesis of this theory in 1907, however, let's consider the following argument. The key idea of general relativity is plain and simple. Think of how we feel in an airplane when it takes off and lands; namely, when it accelerates and decelerates. At both times we feel a force either pulling us back in our seats or pushing us forward. It is as if an invisible power is pushing or pulling us. Pulled by an invisible force? – sounds like gravity (why does an apple fall?), doesn't it? Just as spooky too.

But before there were airplanes there were elevators. The invention of the electric motor was necessary for the rise of the skyscraper in the late-19th century, the time of Einstein's youth. [103] Which brings us to Einstein's second famous thought experiment, Figure **5**.

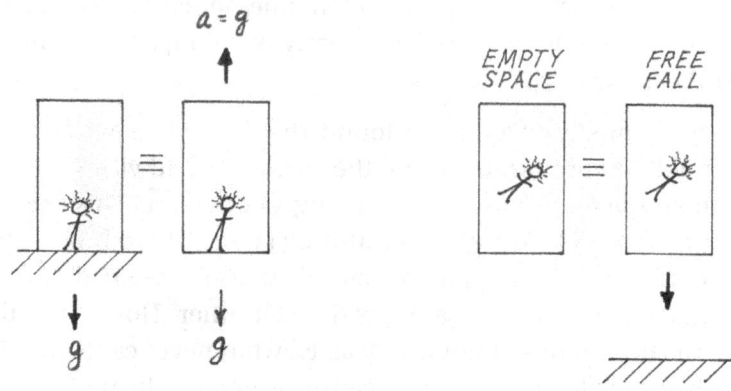

Figure 5
Einstein's 1907 thought experiment on the identity of gravity and acceleration.

On the left is pair of images of the guy with the crazy hair in a box, with no information about the outside world. The far left image of the left pair shows him in the box on the ground, with gravity pulling him (and anything with him) downward by force given by **g**. Whereas the right image of the pair shows him somewhere in outer space, but being accelerated upward with the power of **g**, as in an elevator. In the case he also experiences a force downward, identical to the case of being on the ground. The three parallel lines is the logical symbol of identity (taken from the two parallel lines of the equal symbol). It all means that the guy in the box cannot distinguish between the two experiences. No matter what experiments he performs in the box, he can never know if he is really on the earth or really out in space. The experiences are identical.

The pair on images on the right, carries this idea further. Indeed, this case was probably the source of the thought experiment of 1907 in Einstein's mind. Let me explain. The left image of the right pair shows our guy in the box floating in empty space. The right image has him near the earth and falling in what is called free fall, as gravity pulls the box toward the ground. As for the left pair, so for this right pair: the experiences of the guy in the box are identical. There is no way for him to distinguish between floating in empty space and falling toward the earth, although surely the latter does not last indefinitely. Supposedly, Einstein's key insight that led to all this was his realization that for a person in free fall, it is as if there is no gravity acting on him. Hence the identification of the right pair. He tells us about this in a 1920 document, in which he refers to this idea as "the happiest thought" of his life.[104]

Another way of interpreting this is to say that it is as if you can turn off the gravitational force, just as electrical and magnetic forces can be turned on and off (think of an electromagnet). This may also have led to his later quest to unite gravity and electromagnetism, what he called his unified field theory. (More on that, below.) For now, Einstein's quest starting in 1907 was to bring gravity into relativity theory as a form of acceleration, nothing more.

But how to work this out in a formal mathematical framework, like other theories in physics? – that was the problem. It turned out to be a formidable task, for it required a branch of mathematics that he had never been taught.

He was deeply into this task when the family moved from Prague back to Zürich, a move that proved to be auspicious. From his idea of the modification of space and time in the 1905 formulation of relativity, which was then expanded to four dimensions in 1908 by Minkowski (as mentioned in Chapter II), Einstein conceived of gravity as being caused by the curvature of that 4-D space. That is, space itself, being in constant contact with all bodies of matter, wraps or warps around matter, and it is this bending of space that we experience as a spooky attractive force.

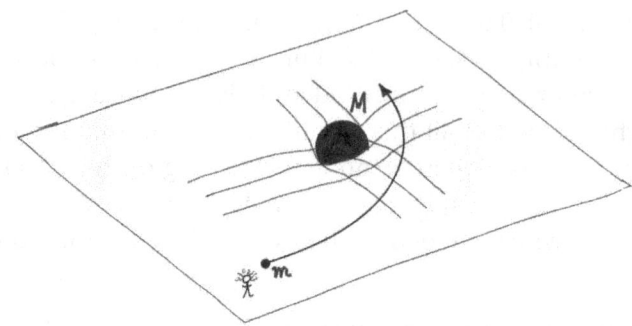

Figure 6
The 3-D analogy of Einstein's 4-D model of gravity.

I visualize this by the 2-D analogy of Figure **6**. Our 2-D guy does not realize that what he sees as an attractive force between masses **m** and **M**, is really due to the matter bending the 3-D space around the mass. That is, the 2-D person sees an attraction between two bodies of matter (**m** & **M**), because of the curved path of the thrown object (**m**) as it approaches **M**. But it is really caused by body **M** curving, warping, or bending the space around it into the 3rd dimension. Warped space is experienced, or behaves, as if there were an invisible force, attracting the moving object (**m**). There is no "spooky" force of attraction. In short, Newton's mysterious attraction was replaced by Einstein's bent space. Therefore, by analogy, for 3-D people, like us, space is bent into the 4th dimension.

This was a clear and simple visual idea, but for Einstein to make it work he needed a mathematical grounding in geometry, since it meant that what we call a gravitational force really is due to 4-D space twisted around matter. After all, if gravity is being explained by the curvature of space, then we are ultimately talking about geometry. But it was not an easy task, for it required the use of a rather new form of geometry developed in the late-19th century called non-Euclidean geometry (that is, beyond the "holy" Euclid of Einstein's youth). When added to calculus it was called tensor calculus, and there was not even a published textbook for it.

This brings us back to the auspicious nature of the move back to the Poly in Zürich. Although Minkowski was gone, Einstein's close friend Grossmann was teaching math there, and importantly he was knowledgeable in the tensor calculus needed to explain gravity by geometry. In fact, non-Euclidean geometry was the subject of his PhD thesis, which Einstein once mentioned in a letter to Mileva, making the candid remark: "I don't know exactly what it is." [105] Well, he eventually found out what it was all about.

So Grossmann, for a third time, came to Einstein's rescue at a critical moment in his life, as he taught his old friend this new form of geometry. Here was a situation in which even Einstein needed help; or put as a historical question: "Would we have our *Einstein* today without Grossmann?" The result was a series of papers co-authored by Grossmann and Einstein as they developed a new theory of gravity. Once Einstein got a handle on this, he was on his own. He was still working on this problem when he moved to Berlin.

As World War I went on around him, he lost himself in this task of wrestling with the tensor calculus equations and trying to coax out gravity through many pages of calculations. During the month of November 1915, he delivered four lectures to his Prussian Academy colleagues, putting forth his quest to solve this puzzle while he was *still* working it out. It was a gutsy thing to do. Such chutzpah! What if it failed? He would be exposed as a failure in front of his peers. By the third lecture, he was able to present some success. The planet Mercury had a very small, but real, extra motion, discovered in the mid-19th century, which could not be explained by Newton's laws alone, and was an unsolved problem at the time. Einstein was able to account for this extra motion by his equations. This deduction was a clue that he was on the right track. He wrote elatedly to a friend: "I was beside myself for days with joyous excitement," and he further revealed that it gave him palpitations of the heart.

But he was not finished: he went on at the end of the lecture with a prediction from his equations. He predicted that a beam of light would be bent by gravity due to the curvature of space. Actually he had deduced this bending in 1911, but now the number he got was twice the one he formerly arrived at. (As seen, this was later verified in the 1919 solar eclipse experiment.) This spurred him on so that finally, in the fourth lecture, he was able to put forth his final formulation of a tensor equation explaining gravity by a non-Euclidean geometry that bends space into a 4th dimension. To repeat: a plain way of picturing this, as seen with our 2-D/3-D analogy, is Figure **6**. Notice how this 2-D guy having a bad hair day infers that a force from mass **M** explains the attraction of the thrown object (**m**) curving around it, whereas it is really due to the bending of space itself by **M**. By analogy, we postulate that in our 3-D world, what we call the attraction of gravity is really caused by the curvature of space by matter into the 4th dimension. The math was formidable, but the visualization (Figure **6**) is simple.

In his escape from the ravages of war into the repose of mathematical physics, Einstein brought to fruition a task he spent almost a decade on. As quoted in Chapter **II**, in 1933 he recalled the task as "years of anxious searching in the dark, with their intense longing, their alternations of confidence and exhaustion and the final emergence into the light." [106] No wonder he got palpitations of the heart.

These months of hard work finalizing the equations of general relativity at the end of 1915, were followed early in 1916 by a long, landmark paper summarizing, and teaching almost textbook-like, the mathematical fundamentals of non-Euclidean geometry and tensor calculus, followed with the derivation of the 4-D equation explaining gravity by the curvature of space. This was another culmination point. Today the paper is a masterpiece of physics.

Yet, writing one the greatest papers in the history of physics was not enough. He still did not stop, for he put his ideas in a popular book, expounding relativity to the intelligent layperson. He told Besso he needed to write this book – to make clear how relativity is ultimately simple, although seemly very difficult.[107] The book, *Relativity: The Special and the General Theory*, which he finished in December 1916, is still in print. [108]

As I write this, the best *scientific* biography of Einstein is still that of Abraham Pais, *Subtle is the Lord* (1982). For this monumental work, Pais had to slog through hundreds of publications, especially scientific ones. One aspect of Einstein's scientific life that kept forcing itself on Pais as he worked on the manuscript was the relentlessness with which Einstein kept doggedly working myriad scientific problems at the frontier of physics, ever revealing an active, fertile, and original mind from, at least, 1905 through the mid-1920s. (Pais's book alone gives the lie to those bigots who claim that Einstein was a plagiarist.) At one point in his book, Pais is compelled to pause and exclaim: "Does the man never stop [thinking/working]?" [109] A rhetorical question, to be sure. Both Pais and we know the answer is "No." The man had guts, perseverance, resolve - in a (Yiddish) word, chutzpah.

The popular book on relativity was followed by the cosmological model of 1917, and this brings us to where we began this chapter. From a visual (geometrical) viewpoint, the 1917 model therefore took Figure **6** (the local curvature of space by gravity) and summed-it-up throughout the entire universe, such that all of space was curved back into itself, resulting in the 3-D sphere in Figure **2** being a reflection of the 4-D universe. Don't forget: this is a simple visual way of bypassing lots and lots of tightly packed tensor calculus equations.

At the time, Einstein assumed that the universe was static, which was the most reasonable conjecture based on contemporary astronomical information. However, his final equation for the universe according to general relativity predicted a non-static model. Taking a rather unadventurous approach – not a common thing for him to do, surely – he added an extra term to his equation to stabilize the model, and he called that term the cosmological constant. Of course, as said, this was reasonable, given the data from the astronomers, for otherwise what would an unstable universe look like? In fact, he initially believed that this constant was another of the fundamental constants in nature, such as the speed of light. This 1917 conjecture, however, did not turn out to be a discovery, as he thought.

Over the next decade, Einstein was challenged about his model three times. In each case, there was an alternative interpretation of his equation without the cosmological constant, and each recognized the possibility that the universe is not stable – for example, that it would somehow expand. Einstein rejected each, holding to the evidence from astronomy of only a minimal movement among the stars.

The most important challenge was from the Belgian physicist and Jesuit priest, Georges Lemaître, in a paper published in 1927. He showed that beginning with Einstein's static model, and taking away the cosmological constant, the universe would continually expand outward. It was, in short, a serious prediction of an expanding universe based on Einstein's theory. Interestingly, and importantly, about six months after Lemaître published his paper, he met Einstein at a conference in Brussels. Einstein, however, was not aware of the paper, since it appeared in an obscure journal. Nonetheless, upon having the model explained to him, Einstein is quoted as responding to Lemaître with this declaration: "Your calculations are correct, but your physical insight is abominable." [110] It was not an unreasonable response at the time. Consider the alterative: look again at Figure **2**. An expanding model would mean that this sphere is like a balloon that is being blown up, with the result that all stars are receding from each other at any point in the space of this universe; and, moreover, that space itself is continually being created as the matter (stars) expands into it. It was, in a fundamental sense, a logical extension of the malleability of space as entailed in Einstein's general relativity theory. Yet, to make a pun, it was a stretch of the imagination. And, it seems, even Einstein's imagination was limited in how far he would go. [111] This intransigence would change, however, sometime from mid-1930 to the early months of 1931.

Prior to Einstein's first trip to Caltech, Edwin Hubble made two monumental discoveries in the 1920s. Hubble was an astronomer at Mt. Wilson Observatory near Caltech, with the largest telescope in the world at the time. First, he found that there were other galaxies (he called them nebulae) external to our Milky Way, an assertion that previously was just a fantasy according to most astronomers. By the late-1920s, therefore, the universe was conceived of as containing numerous galaxies extending throughout space. (In Figure **2,** think of replacing all the stars with galaxies.) Next, Hubble announced his second discovery: almost all of these objects seemed to be expanding outward in all directions, since they were exhibiting what we call today Doppler redshift. When objects move away from us, their light is shifted to the red part of the spectrum; alternatively, if coming closer, it shifts toward the blue. Christian Doppler discovered this fact in the mid-19th century. Since almost all galaxies were showing redshift, this opened up the possibility that Lemaître's universe was true. Hubble balked at this interpretation throughout his life, but others did not. [112] After some initial reluctance, many scientists by the early 1930s took some version of the expanding universe seriously.

And Einstein? Well, in the summer of 1930, he received an honorary degree from Cambridge University, and there he spent some time with the astronomer Eddington, who (recall) had been instrumental in the famous solar eclipse experiment in 1919. Importantly, Eddington was also an expert in relativity, having written a book on the subject in the 1920s, and was very familiar with the work of Lemaître and Hubble. Indeed, in 1931 he had Lemaître's paper lifted from obscurity by reprinting it through the Royal Society. [113] There is, therefore, significant reason to believe that Einstein's transformation from a static to an expanding model for his cosmology was set in motion during this interaction with Eddington. [114] So, by the time he arrived at Caltech in December, he was prepared to assert, as reported in the press early in 1931, that he abandoned his static model of the universe; he said that his old static idea was "smashed ... like a hammer blow" due to the discovery of the redshift of the distinct galaxies. And he was swinging his arm while declaring this. [115] From then on, the static model was never again taken seriously, as he discarded the cosmological constant. When the topic arose in the early 1950s, for example, he is quoted as exclaiming: "[positing the cosmological constant was] the biggest blunder of my life."[116] I assume he was speaking of his scientific life, not his personal one.

The moral of this story? Even Einstein could be stubborn for the wrong reason, displaying a less than imaginative view of the world. Too bad. Think of it: he could have predicted the expansion of the universe! As if he did not do enough already, eh? In the end, at least, Einstein's model became the basis of all cosmic models for the rest of the century – and, so far, into the next millennium. Not bad.

For all this extraordinary work, as seen, Einstein was awarded the 1921 Nobel Prize for Physics. Because he was on the trip to the Far East when the Prize was announced in 1922, he did not deliver his Nobel lecture until 1923. There are several myths around Einstein and the Prize. Here is a brief outline of what happened.

It is commonly said that he first heard that he won when he received a cable announcing the Prize during his voyage to Japan, but actually he knew he won before he left Europe. The cable came as no surprise. It is also said that the Prize was given for his work on the photoelectric effect because relativity was still too radical at the time. In fact, the photoelectric effect was just as radical. As seen, even though Einstein's equation for the effect was proven true experientially, the theory of the photon particle was rejected by almost all physicists (such as Bohr), except Einstein. As a result, the Nobel citation began with relativity, next it mentioned the application to cosmology, then his work supporting atomism (such as Brownian motion), and in the longest part discussed the work on the *equation* for the photoelectric effect. In essence, the Prize was about the miracle year (1905) in a nutshell, plus the 1917 cosmology paper.

Finally, the conventional story is that when Einstein delivered his acceptance speech, he purposely ignored the photoelectric effect and insolently discussed relativity – another clear example of chutzpah. Yet this is not quite true. The Nobel committee actually suggested that he speak on relativity, and he agreed. So, the speech began, as requested, with the special and general theories of relativity, but as he approached the end Einstein brought up a "subject of lively interest" – namely, his quest for what he called a "unified field theory," which was what he intended to speak about in the first place. So, Einstein, the contrarian, at least partially, got his way. [117] And this brings us to the topic of the rest of this chapter – what Einstein did for the remainder of his scientific life.

As mentioned in Chapter II, after those years of constant arduous work that culminated in general relativity, Einstein came down with various ailments, and Elsa nursed him back to health, after which they were married. Recovering from this period of illness, he got back to his real love, and continued making important contributions to quantum physics into the mid-1920s. A very brief survey of this would start with the 1905 quantum of light (photon) theory, in 1906 the application of quantum effects to the specific heats of solids, 1909 and the idea of a wave-particle duality for light and all electromagnetism, 1917 and the calculation of a momentum for the photon, 1923 on the relativistic energy-momentum of the photon, and finally quantum statistics worked out with the Bengali physicist Satyendra Nath Bose in 1924-25, now called Bose-Einstein statistics. It was a momentous achievement.

In the meantime, however, he became enamored with another foundational problem. First special and then general relativity had exhibited a progression from a restricted theory toward a more comprehensive theory. Adding the 1917 cosmology idea made it encompass the entire universe. But not quite. It surely explained the spooky idea of gravity as a mysterious force in nature. However, there was another force in nature in the 1920s that was not explained by relativity: the mysterious power of electromagnetism. Both electricity and magnetism had separate, and parallel, histories from ancient times until the 19th century, when it was discovered that they were two parts of the same thing – electromagnetism. Moving magnets could produce electricity, as in a dynamo, and electricity in a coil of wire could become an electromagnet. These facts Einstein learned about as a child in his father's business. This was the technological side of electromagnetism, which he then dealt with further in his job at the patent office.

The scientific side of the story of electromagnetism he learned at university, which he supplemented with self-study. Two key 19th century scientists in this development were the Englishman, Michael Faraday, and the Scotsman, James Clerk Maxwell. Faraday introduced the key concept (and even the word, used in a scientific context) of a "field" of force. Lacking sufficient schooling in mathematics, Faraday used the mental images of electric and magnetic fields, and was able to predict and explain how magnetism is converted into electricity, and vice versa. (Later the image was applied to gravity, but it was not part of Newton's original conception.) Faraday's work was followed by the work of Maxwell, who converted Faraday's conceptualizations into mathematical expressions that eventually became a quartet of equations known as Maxwell's equations, which (as noted in Chapter II) were not taught by Einstein's Poly teachers.

Einstein's relativity theory was grounded on this work; indeed, in some ways the field image of (say) gravity was a stepping-stone to the ability of space to move matter. Faraday believed his field to be a real entity, and Einstein conceived of space as *not* nothing (no-thing) but as a malleable real thing. Thinking visually, it was then a short conceptual jump from the image of the gravitational field to the warped space constituting that field. It is of more than passing interest to note that in his study in Berlin and again in his study in Princeton, Einstein had portraits of only these scientists hanging on the wall: Newton, Faraday, and Maxwell.

The task Einstein set for himself in the 1920s was to subsume the electromagnetic force into general relativity. Put plainly, just as he deduced Newton's gravitation force from the relativity equation, he hoped to deduce the electromagnetic force too. From a geometrical viewpoint: he had explained gravity by a warping of space (in non-Euclidean geometry); now he wanted to modify that geometry further in order to generate electrical and magnetic fields. This sort of bigger equation would contain the modifications of space to generate all the forces in nature. In this way, one comprehensive equation of space would encompass both forces in nature, electromagnetic and gravitational, since ultimately both fields were caused by different warpings of the same space. For this reason, he called it a unified field theory.

Initially, in the 1920s, several scientists went along for the ride, working with or without Einstein in a quest to find this master equation of the universe. But the task proved much more difficult than originally thought, and one scientist after another dropped out. By the mid-1930s, when Einstein was permanently in Princeton, almost he alone (along with various assistants who came and went) pursued the unification quest. By the 1940s he was viewed as a stubborn old man, wasting his time on the impossible task of unifying the forces of nature. A young J. Robert Oppenheimer, who would later be his boss at the Institute, visited the Institute as a very young physicist and said in a letter that Einstein was "completely cuckoo." [118]

Einstein did not mind being shunned by his colleagues. He knew he was chasing a chimera. But he also knew that few physicists would take the time to do so, since they could not afford, as he could, to toil for years on possibly an unproductive task.

There was, however, another reason for his isolation, which went back to the late-1920s. As noted before, he made many of the major contributions to quantum physics from the start, and this continued (sometimes alone, or with collaborators) into the mid-1920s, when finally everyone else accepted the reality of the photon. Yet, peculiarly, just as Einstein's photon became part of quantum theory, Einstein quit believing in the quantum theory. This, of course, I need to explain.

In the mid-1920s there arose an interpretation of quantum physics that Einstein rejected. It was put forth primarily by his close friend, the Dane Niels Bohr, famous for the modern model of the atom, and for which he was awarded the Nobel Prize for 1922. The squabble between them, mainly in public at conferences and in private interactions, has been dubbed by historians as the Bohr-Einstein debate. The discussions, of course, involved very heavy and complicated quantum physics stuff – but, I submit, there is a very basic and clear way of explaining this debate to the intelligent reader. Here goes.

A critical (maybe *the* critical) topic around which the debate circled is the role of probability and statistics in modern physics. Einstein made important contributions to the use of statistics in physics even before 1905. Recall that much of this work was around the idea of atoms, lots and lots of atoms in the microscopic world, and to deal with so many things at once, statistics was used. Statistics hence became central to the quantum world. But what does it mean to apply this mathematical tool to nature? Well, in the macroscopic world, such as lots of people, we apply this tool because we cannot predict how any one person will behave, but we can deduce the probability that a group of them will behave in certain ways (think of actuarial tables used by insurance companies). That is how Einstein interpreted statistics in the micro-world. It was about our limitations in knowing the details; we can only speak of probabilities in what we can *know* about the atomic (and subatomic) world.

Bohr, however, made a radical leap: to him the probabilities were not due to our limitations in the process of *knowing* about the micro-world, but were about the world *itself*. There were no real discrete entities in that world; the probabilities used in quantum physics were of the world *itself*. Nature itself was statistical, he proclaimed. It was as if in the macroscopic world there were no real people, only the probability of some people.

Einstein would have none of that. Recall the sentence from his autobiography that I quoted in Chapter **I**, about his liberation from the religious period in his pre-teen life. "Beyond the self there is the vast world, which exists independently of human beings, and that stands before us like a great, eternal riddle, at least partially accessible to our inspection and thinking." He experienced the revelation that beyond the selfish solipsism of his youth there was a real external world that was knowable, at least partially - the limitation being due to the statistical nature of our *knowledge* of the world, not the world *itself*. In an unpublished letter, written about the time he was working on his autobiography, he wrote: "In truth, I never believed that the foundation of [quantum] physics could consist of laws of a statistical nature." [119] Said another way: the fact that quantum theory must rely upon statistics to work, means that the theory is incomplete. The problem was with the theory, not the world.

By the 1930s Bohr's interpretation prevailed, as a new generation of physicists embraced this radical view of the strange quantum world. Since he put forth this idea from his Institute in Copenhagen, Denmark, it became known as the "Copenhagen interpretation of quantum physics" and it was dogma in the textbooks I used as a physics student in the 1960s.

But Einstein never embraced it. Listen to what he said to a fellow physicist in a letter in 1944: "You believe in a God who plays dice, whereas I believe in perfect laws in a world of existing things, in so far as they are real, which I try to understand with wild speculation." [120] For his continued resistance to the statistical quantum theory, Einstein's fellow physicists ostracized him to end of his life.

Although the popular press still saw him as the epitome of what a physicist is, to his physics colleagues he was, as he said himself, an old fossil. Did he also identify with the excommunicated Spinoza, his favorite philosopher? I would not be surprised.

There may be a further factor that contributed to his isolation. When Einstein crossed the Atlantic in 1933, never to return, he transferred his allegiance from Europe to the USA. He metaphorically crossed a linguistic ocean, for he never was fully fluent in English, always speaking with a thick German accent, and not completely comfortable. This alienated him further from his fellow scientists, who were already castigating him for his stubborn rejection of quantum physics. Nonetheless, as noted, this pariah status among scientists was in contrast to his stature in the public eye as the greatest scientist since Newton, the smartest man in the world, and a sage on social, political, and moral matters. [121]

There is one more reason that Einstein worked unflinchingly on the unified field theory. It is not commonly known but as he strived toward a solution to this problem, he became convinced that if he derived this master equation, uniting the gravitational field with the electromagnetic field, then the fundamental equations of quantum physics would, so to speak, fall out too. This would mean that the final relativity theory would not only merge gravity with electromagnetism, but quantum physics as well. Then quantum physics would be complete, as it was subsumed within a grand relativity theory. The pages of calculations found on his deathbed were his final effort toward that light in the distance.

Perhaps the best vindication today of Einstein's late years as a pariah is the fact that the best minds in physics are pursuing – guess what? – a theory of everything, as they call it. Ironically, as I write this, the physicists at the very Institute where "Einstein the pariah" spent most of his time on this problem, are together diligently plugging away at unifying all the forces of nature. What was seen as chasing a chimera is now middle-of-the-road physics. I imagine Einstein smiling from above. Perhaps smiling with that slightly impish smirk found in Photo **1**. At least, I would like to think so. The thought is comforting.

(FYI: At minimum, I should add here that after Einstein commenced his quest to unify electricity and gravity, two other forces in nature were discovered within the nucleus of the atom – the strong and weak nuclear forces. But Einstein ignored this discovery. After he died, physicists were able to develop a theory that unified these two forces with the electrical force. Nevertheless, the final unification of this trilogy of forces with gravity remains a great puzzle, and hence the quest goes on ...)

I believe it is proper, in a chapter called Chutzpah, to bring up two examples of Einstein's stubbornness that did *not* bear fruit – rather like we saw with the cosmological constant. The two examples are: The Einstein-deHaas Effect and Mach's Principle. Moreover, since these two are not well-known in the general literature on Einstein, it seems sensible to introduce them to the general reader here. So, here goes.

In 1915, when Einstein was deeply engaged in trying to complete his general theory of relativity, he strangely took off in another direction: becoming involved in a physical experiment on electromagnetism. Where did he find the time? And why did he do something so seemingly out-of-character for this theoretical physicist?

The undertaking apparently was done as a favor to his close and revered friend, the great Dutch physicist H. A. Lorentz, whose son-in-law W.J. deHaas, was recently granted a temporary position at a laboratory in Berlin and needed some guidance. DeHaas had worked on some experiments on magnetism in Leiden, and the experiment that was devised had relevance to an atomic physics problem of interest to Einstein.

The essence of the experiment can be easily explained by an analogy with a well-known phenomenon. Think of an ordinary garden hose wound up on a reel and connected to a water source. When the water is turned on, there is a sort of kick (really a torque) as the water winds its way around the hose; the kick is a quick rotational reaction of the reel in the direction opposite to the water flow. It is an example of Newton's law of action and reaction. Einstein's idea was that a similar phenomenon should occur when an electric current flows through a coil of wire, producing a magnetic field. Accordingly, they devised an experiment to test this.

The experiment used an iron rod hanging vertically from a thin wire, with the rod suspended within an electric coil. When a current flowed through the coil, the magnetism induced in the rod should make the rod rotate – like the water and the hose. That kick of the rod was what the experimenters were looking for qualitatively. Without going into the mathematics of it all, the quantitative result they wanted was something called the gyromagnetic ratio(g), which Einstein predicted would be g=1. [122]

As noted, for Einstein the experiment had relevance to an atomic physics problem: it was called zero-point energy. The idea came from Planck; namely, that an atom at absolute zero temperature would still possess energy. Einstein believed that an orbiting electron was a model for this zero-point energy. Moreover, he thought that this experiment would also confirm Bohr's model of the atom (1913), where electrons orbit the nucleus without losing their energy, contrary to classical physics.

Although reasonably simple in principle, the execution of the experiment was quite thorny. Nonetheless, Einstein and deHaas persevered and got some results. Einstein wrote to Besso: "With it [i.e., the experiment] the existence of zero-point energy has been proven in a single instance. A wonderful experiment; what a pity that you can't see it. And how treacherous nature is, when you want to deal with it experimentally! Experimenting is becoming a passion for me even in my old age."

It is peculiar that he speaks of his "old age" when he was not quite thirty-six, but not odd that he got engrossed in such laboratory hands-on work. Remember that he spent much of his Poly years cutting classes to spend time in the labs. The experience may also have harkened back to his childhood enjoyment of rummaging through the electrical equipment in his father's business.

With this experiment, Einstein and deHaas got, so to speak, the kick they were looking for, qualitatively. But what about the gyromagnetic ratio? Well, they performed two sets of experiments, obtaining g=1.45 and g=1.02. The latter was close to the predicted value, and so they discarded the first set and published the result of the second only, thus confirming the theory – or so they believed. When the result appeared in 1915, Bohr was delighted, saying it was "direct support" for his atomic model.

Over the next few years, however, subsequent experiments by others did not duplicate their quantitative result. The later numbers obtained were closer to $g=2$ than $g=1$. But Einstein insisted that something was wrong with these subsequent experiments, since his theory predicted $g=1$. So it did, but therein was the problem. By the early 1920s, with the discovery of what is called electron spin, it became clear that the angular momentum (or "spin") of the electrons themselves, not their orbital motion, was the cause of the gyromagnetic ratio, and such spin gave $g=2$. Thus, a different theory fit the later experiments.

In hence another instance, we find Einstein's stubbornness getting the best of him, for if they had not thrown out the first set of data, they may have performed a third set, or more, and eventually been forced to reconsider their theory. But that was not to be: here was another prediction that was thwarted by Einstein's chutzpah.

My second example, on Mach's Principle, is a much more complex story, and is quite different from the straightforward framework of the Einstein-deHaas Effect narrative on the interplay between theory and experiment.

For this second story we need some background. Galileo, as seen, was the source of the idea of the relativity of motion. Furthermore, this relativity of motion was grounded on the concept of what was later called inertia. The word came from the Latin word for inert; namely, the tendency for a body to remain in a given state. As a principle of motion, the law of inertia states that a body in motion tends to stay in that state of motion, unless forced to slow down or stop. Similarly, a body at rest tends to stay at rest unless forced to move. So, for example, when riding in an airplane moving at a constant speed, all things – you, the plane, the things within the plane – obey this law of inertia and hence you experience this ride as if everything were at rest. Einstein called such a situation an inertial system. Only when the plane slows down, speeds up, or banks, do you feel a force; this is then a non-inertial system.

The law was based on the experience of moving bodies, observing how they act. It did not explain *why* these things happened. For some scientists this was a problem, since it was commonly believed that science should explain why things happen, not just describe how they do. Yet, the idea of inertia, especially after Newton used it in his laws of motion that seemed to account for the motion of all bodies in the universe, was taken as a real thing in the world, even without giving it a cause.

This then brings us to Einstein around the turn of the 20th century, when his friend Besso recommended a book: *The Science of Mechanics*, written in the 1880s by Ernst Mach, a distinguished scientist and philosopher. Of the many themes and ideas in that book, the main one that had an impact on Einstein was Mach's critique of Newton's concept of absolute space that had appeared in his masterpiece on the physics of motion, the *Principia* (1687). Although Newton used Galileo's principle of relativity and the law of inertia in his physics of motion, he did not believe that we were trapped in a universe of only relative motion. However, try as he might to find a way of detecting absolute motion for what was later called an inertial system, he could not. But he did conceive of a way out of the conundrum for non-inertial motion. He presented his idea in terms of a concrete example. We call it today Newton's bucket experiment.

Consider a bucket full of water hanging from a rope that one can wind up and release. When released, the bucket rotates, and as it does the water is forced radially out from the center of the bucket. As such, the water splashes more and more as the bucket increases in rotational speed until it slows down and stops. So, what to make of this? Or, better said, what did Newton make of this? The simple answer is that it proved the existence of absolute motion, for to hold to the relativity principle in this case led to an absurdity. Here's why. The relativity principle would assert that the following statement was equally true: the bucket was at rest, and the rest of the world was moving around it. Well, that could be true, but there was a problem: why was the water splashing outward? For Newton this was the crucial distinction between the two cases: for him the splashing water was proof that the bucket was really rotating, and hence he was detecting real, absolute motion. Absolute motion in his mind — and in the mind of God.

The latter was not a trivial deduction. Indeed, he wrote the *Principia* to prove the existence of God. For God was the Absolute of all absolutes, and right near the start of the book he said that he wrote the book to distinguish between relative and absolute motions. The authority of Newton's physics in subsequent centuries carried his bucket idea along as proof of absolute motion. That is, until Mach in the late-19th century. So, how did Mach counter Newton's bucket idea? Well, the straight answer is that he said this: if the bucket is at rest, and the world around it turns, the water still splashes. Thus all motion is relative. The splashing water proves nothing. How could Mach say this?

There has been much controversy and confusion over Mach's idea ever since, but I believe there is a clear and informative way of explaining this that I have not seen anywhere else. [123]

Let's begin in the middle of the 19th century with a discovery by the Frenchman, Leon Foucault. While working with swinging pendulums, he found that over time the plane of a pendulum rotates around a vertical axis going through the fulcrum. Upon timing this rotational motion, he found it to be approximately a day, but little bit less. What did this mean? Indeed, what was the cause of this motion? Well, a day is the time of rotation of the earth, which leads to the question: how could the daily rotation of the earth account for this? Consider a pendulum at the North Pole, swinging back and forth. Any rotational motion would be seen as relative to the fixed stars in the sky. Thus if the viewer experienced the plane of the pendulum rotating, that motion would be due to the earth's rotation with respect to the stars; in other words, with respect to the absolute space that Newton spoke of. Further reflection on this: if the pendulum were on the equator, there would be no rotational motion of the plane of the pendulum. Therefore, the period of the rotational motion is a function of the latitude of the place on the earth where the viewer is. Using this hypothesis, Foucault made a calculation using his position in Paris, and found that it explained the period he measured as being a little less than a day. He went on to demonstrate this motion using a 100-foot pendulum in the Rotunda of the Pantheon in Paris. Today many examples of what we call Foucault pendulums are found in public places throughout the world.

Foucault's discovery was interpreted as proof of Newton's idea of absolute space and motion by almost all scientists – but not Mach. For him it was just another example of relative motion, in this case rotational motion. To easily understand his argument, we need one more thought experiment; this one I made up. It is an historical thought experiment. You are a scientist living in the Middle Ages (probably a monk, unless you are a wealthy noble), and you are doing some experiments with a pendulum. You discover the rotation of the plane of the pendulum and you measure it as being nearly a day. What do you make of this? Firstly, the cosmic model you are using is the ancient geocentric one, with the earth as a sphere at the center of the universe; and a day, on this model, is the time of rotation of the heavens around the earth – sun, moon, stars, etc. You also know that a compass always points north; that the needle on the compass is a magnet; and that magnetism is a mysterious power displaying attraction and repulsion across distances. You therefore assert that the compass points north due to an attraction of the magnetic needle with the north star in the sky, which is always in that fixed position. How could it be otherwise? With all this information at your command, you deduce, by analogy, that the rotation of the pendulum is due to an attraction between the pendulum bob and the stars. The bob is just following the daily motion of all of the heavens. All this is merely further confirmation of the connection between the world above and the world below, a belief that forms the basis of astrology, a popular pastime claiming to predict the future.

If we now scoot forward many centuries to Mach: he is working in the late-19th century with the earth rotating beneath him in a day. So, if he does not believe in absolute space (which he calls "monstrous"), nor in any corresponding absolute motion, then how can he interpret what he sees? Like our Medieval scientist, he has only one answer: the stars. The bob is attracted to the fixed stars, an idea not very farfetched in the 1800s, since Newton had put forth that gravity acted among all bodies of matter in the universe, and at extreme distances. Accordingly, to pull this all together, the water in Newton's bucket would still splash outward, even if the bucket were at rest on the earth, and the heavens rotated around it. This attraction between the bob and the stars explained the fixed positon of the pendulum's plane with the stars (avoiding the postulation of absolute space) – and, as a bonus, explained why the plane exhibits inertia (remaining at rest) with respect to the rotation of the earth beneath it. In short, Mach found a causal explanation of inertia.

Thus we can now bring Einstein back into this story. As seen, starting in 1907, he diligently was searching for a way of bringing gravity into the theory of relativity. As he worked on the problem over subsequent years, he came to the realization that Mach's idea had a role to play in his theory. Since all motion will be relative motion, then there is no absolute space, as Newton proposed. The bucket experiment, therefore, must be interpreted in terms of relative motion, and this meant that Mach's Principle, as he called it, would explain how this could be true.

It was a very seductive idea for him. It helped him understand the relativity of linear and non-linear motion; or integrating constant speed and acceleration into one; or combining inertial and non-inertial motion; or, even more, granting a *cause* to the real thing called inertia. He first published this in 1912, writing that "the *entire* inertia of a mass point is an effect of the presence of all other masses [in the universe], which is based on a kind of interaction with the latter." A year later he wrote to Mach himself on the topic. He said that if his (Einstein's) idea of explaining gravity by relative motion is correct, then "it follows of necessity that *inertia* has its origin in some kind of interaction of the bodies, exactly in accordance with your [Mach's] argument about Newton's bucket experiment." [124] I believe that all this reinforced his belief that this quest to bring gravity into relativity was not futile.

Some other scientists took up the torch for Mach's Principle for a while, but most eventually dropped it for various reasons. Einstein, however, kept the torch lit, and indeed even brightened it, when he developed his cosmological model in 1917. Recall that this model was a static and finite image of the universe, as in Figure **2**. The idea of a mutual attraction among all bodies of matter interacting with each other across a finite space, and balanced by the repulsion of the cosmological constant, was a simple and clear image to comprehend. So Einstein apparently held onto the idea throughout the 1920s (but, see below).

As seen, by the early 1930s, especially after the Caltech visits, he was forced to abandon the static model and embrace the expanding universe, which, it seemed, was not compatible with Mach's Principle. After all, without the fixed stars out there in space, there was no "other matter" to provide the mutual attraction that explained inertia. What "other matter" there was, was flying away into deeper and deeper space that was continually being created. For this reason, Einstein abandoned the Principle, at least by the early 1930s.

There is, however, one more path to follow in this story. I am convinced that despite the deductive aspect of Mach's Principle, especially for explaining inertia, Einstein saw a fundamental paradox in using it. The paradox was this: the positive feature of the Principle was that it gave an answer to the cause of inertia; the negative one was, using a term introduced before in this chapter, that it entailed an action-at-a-distance idea of gravity. Recall that one important deduction of general relativity was that curved space was the cause of gravity, and this eliminated gravity as spooky and acting-at-a-distance. So why bring it back through the Principle? Was an answer to the cause of inertia worth it? It seems that for Einstein, at first, the answer was, yes, or maybe – at least we see him still grappling with Mach's Principle around 1930, before his final rejection after the Caltech visits. In his famous lecture at the University of Glasgow in 1933 (quoted above), he admitted that his fascination with Mach's Principle was principally due to its explanation of inertia and was therefore difficult to give up. But, in the end, he did.

And the loss is still with us. As Einstein's biographer Pais wrote: "the origin of inertia is and remains *the* most obscure subject in the theory of particles and fields." [125] A puzzle for present and future physicists, it surely is.

In this chapter on Einstein's science I present the following brief addendum to the revised edition of this book, made necessary because of an experimental discovery announced on February 11, 2016. The knowledgeable reader may guess what this is: for the presses were chock-full of Einsteinian news about the detection of gravitational waves, a prediction from his theory of relativity that was finally confirmed.

The idea of gravitational waves nicely follows from the role of space in his theory of general relativity. Since, in this theory, space is a thing (not nothing) then a body of matter in space is rather like a body in water. Disturb (such as by vibrating) the body, and waves are sent out in circles. Similarly, Einstein predicted that an enormous disturbance to a body of matter (such as a massive explosion) should produce gravitational waves in space. He was not the first to conceive of waves of gravity, which has a prehistory, but he was the first to give the idea a mathematical foundation. He did this in 1916 as part of his completion of his general theory, although he had to make a correction to the math two years later. The century gap between the conception and the confirmation of gravitational waves gives the story a nice symmetry (1916-2016).

But, like most seemingly clear-cut stories, there is more to it. The mathematical calculations of 1916 & 1918 were approximations. So, in the 1930s, Einstein (along with an assistant) attempted an exact solution. The details of this are intriguing – and the eager-beaver reader may wish to track it down [126] – but the relevant point here is this: by this time he doubted whether gravitational waves could even be produced by distant stars. And, as far as we know, Einstein died still doubting their actual existence.

Nonetheless, when they were found in 2016, it was called a confirmation of his prediction. Indeed, instruments at two Laser Interferometer Gravitational-Wave Observatories (LIGO) in Washington State and Louisiana detected the ripples from an explosion produced by two black holes colliding 1.3 billion light-years away. The massive explosion was enough to produce the predicted waves which the very, very sensitive LIGO instruments were able to measure.

Finally, each chapter of this book begins with a brief autobiographical reminiscence relevant to the topic covered. Here, I am ending with one too.

During my many years as a physics and math student, and in an undergraduate class in mechanics, I had an insight that I still remember. It was not a unique or creative discovery, but it had personal significance – for to me it transcended (or abstracted from) the mundane world of just doing physics problems.

I do not recall the specific problem at hand at the time, but it went something like this: if you have a ball of a given weight, which is projected at a given angle at a given speed, how high will it go and how far will it travel? Using the correct equations and manipulating them properly, the resulting numbers can be verified by doing the experiment. The numbers generated by the mathematical equation match the real world (approximately).

Somewhere, in course of solving this problem, it struck me how really extraordinary what I was doing was. I had started with a real world situation that I could duplicate (a ball of a certain weight thrown at a certain angle, etc.) and then I retreated into a world of mathematical symbols and abstractions that followed rules whose only stricture was the canon of logic. I then re-emerged back into the real world with some numbers, which, if checked corresponded to what the ball would really do in that world – within, that is, some small errors, due to friction and other real world departures from the idealized world on paper (or blackboard, or whatever). Say, the ball actually went 40.6 feet, whereas I predicted 41 feet. The small difference was irrelevant; but the closeness was the extraordinary thing. How can this be? How can this abstract world have a relationship to reality? How can what I am doing when I manipulate these mathematical symbols have any bearing on what is happening with objects in the physical world? What did all this mean? As I interpreted this process, it meant that the real world was controlled by the same logic as mathematics. There were underlying powers in the world that obeyed the rules of logic. How incredible that was, I thought.

Little did I know that many scientists had the same experience sometime in their life, and that someday I would be pondering this very topic in the mind of that smart guy, Einstein. To get to my final point on this, let me begin with a brief review of a key thread from the last two chapters.

In his preteen years, Einstein made a break from religion and embraced science in its place. But he did not then see science and religion as entwined in an eternal conflict, as so many of his generation did. Rather, he saw both as embodying truths, but each as of a different kind. Science was about facts, religion about values: the *is* and *ought* of life, respectively. From this evolved his personal theology based on the pantheism of Spinoza, that God "reveals himself in the harmony of all that exists," and his increasing affinity with Judaism as a cultural and ethical collective.[127]

From this we move to another source of the harmony in the external world; namely, his increasing application and appreciation of the role of mathematics. As a student, he viewed math a tool for explaining nature, a means to an end. That was how he used it in special relativity. But with the development of general relativity, and the more abstract world of a 4th dimension and non-Euclidean geometry, he realized that math itself was a manifestation of the creative act, especially as he grappled with trying to unite gravity with electricity. He acknowledged this in the lecture at Oxford in 1933, when he affirmed that "the creative principle resides in mathematics. In a certain sense, therefore, I hold it true that pure thought can grasp reality, as the ancients dreamed." [128] What an incredible thing for a physicist to say (especially one who cut his math classes) – "pure thought can grasp reality." Wow.

It is worth repeating: "pure thought can grasp reality." No wonder he had the gumption to pursue his unified field theory dream to the end of his life.

V. The Hair

In days of yore (in what are often called the 60s, but for me really was the 70s) when I had moderately long hair and a moustache, if I were invited to a costume party, I would always comb my hair and moustache with white powder, mess it up, and don a loose T-shirt and baggy pants. As I walked into the party someone, upon seeing me, would exclaim, "Einstein." The canonical image was, and still is, ubiquitous everywhere.

How and why did Einstein acquire this visage? The answer begins with a short biography of Einstein's hair, starting with the facial hair. Einstein was clean-shaven until around the Aarau year (1895-1896), when he grew a moustache, which he kept for the rest of his life.

Now, the hair on his head. Einstein had very curly hair. As a young man he wore it full and not closely cropped, but not unruly, either – at least, most of the time. See the Aarau picture (Photo **3**): note that his hair is not slicked-back, as are the other men's hair. Around 1920 his hair began to gray, possibly because of the illness of 1918-1919, and the graying continued throughout the 1920s. Occasionally it was wind-blown, but still it was not too long. Then around 1930, he started leaving his hair as a jumbled mess, by not combing it (Photo **4**). He also stopped cutting it regularly, as it was becoming almost completely gray.

Photo 4
Einstein circa 1930. Credit: Universal Images Group / Art Resource, NY.

By the mid-1930s he let it grow longer still and it morphed into the "mad scientist" look that became the visual prototype for drawings, caricatures, and cartoons. Occasionally Elsa would give it a trim, and after she died, Dukas and various "girl friends" were his barbers. Otherwise, he kept this "Einstein" look until he died.

Why did he let his hair go? The short answer, which is no surprise if you have read all of this book so far, is one word: nonconformity. It further links us to the first picture in this book (Photo **1**), where the laid-back contrarian was the only boy smiling in the school class picture.

V. The Hair

But the more robust answer goes something like this. Einstein's contrarian behavior was expressed in his atypical and bohemian social behavior up his divorce from Mileva. In a letter to Elsa in August 1914 he spoke of his fear of marriage to her from the viewpoint of this lifestyle. "Is it a fear of the comfortable life, of nice furniture, of the odium that I [would?] burden myself with, or even of becoming some sort of contented bourgeois?" Horror of horrors, he would become a card-carrying member of the middle class if he married her. Spoken like a true contrarian or nonconformist – or free spirit, eccentric, maverick, what have you. Yes, Elsa was a thoroughly bourgeoisie who enjoyed socializing with well-placed people. Oh, what a philistine she was.

So, anyway, in 1919 Einstein made the plunge, and he sat in the "nice furniture" – for, unless one is a card-carrying masochist, what's not-to-like about a "comfortable life"? Elsa, of course, made the effort to mold her husband's habits into following some rule of social decorum, with clean clothes, proper dress for each occasion, and combed hair. It was not an easy job. She was at first partially successful, but by the late-1920s she threw-in-the-towel, just as she gave up trying to dissuade him from having affairs with other women. Her defeat was expressed in his hair, which often went in several directions at once. Moreover, by then she was so weary of the struggle that she got lazy herself with her own garb. By the early 1930s, she often let her own hair go too – and oddly (or amusingly), they both started looking alike.[129] As always, Albert won, and she joined him.

(BTW: the same thing happened in his relationship with his sister after she moved into his home in Princeton. Her hair grew long and shaggy and in their latter years, they appeared almost as twins.[130])

At times Einstein's hair looked like a halo around his head, as seen in Photo **5**. Saint Einstein? In fact, he once noted, in a self-mocking way, that he was sometimes treated as a Jewish saint. We know now, of course, that he was no saint. His behavior toward women was conventional by the standards of his day, but (at times) abusive by today's rules. Nonetheless, as he used his celebrity status, starting in the 1920s, to put forth his ideas about social issues, and supported various causes, he revealed a strong social conscience to which he was always willing to lend a hand as well as his name.

Photo 5
Einstein circa 1940, with a halo of hair. Credit: Culver Pictures / The Art Archive at Art Resource, NY.

Recall his signing the counter-manifesto in 1914, which was a gutsy thing to do in jingoistic Germany. Later (as seen), he stood up for Eastern European Jews, which was not a major cause even for German Jews, who were often embarrassed by the looks and behavior of those from the East – which shows how integrated these German Jews were into German society. (Well, alas, I should rather say, how integrated they *thought* they were; Hitler and his henchmen would prove them wrong.) We also saw Einstein's concern for the discrimination against African-Americans when he moved to the USA. In the early 1950s, moreover, he was outspoken in his vehement opposition to the "witch hunt" trials of Senator Joseph McCarthy, which he called an Inquisition. Einstein urged witnesses to refuse to answer questions to McCarthy's committee and the other Congressional committees searching for Communists in America, and he did this openly, as reported in the press. It one point he blurted out to a friend, "Do you think they will arrest me"? The authorities did not; as well, he was never called to testify to any committee, although the thought did cross his mind too.

Einstein realized that this obsession with Communism was a peculiar American foible. It stemmed, in part, from a lack of understanding of socialism, due to the fixation with the role of individual freedom in society. He wrote about this in an important essay called "Why Socialism?" in 1949.[131] In it he spoke of the "economic anarchy of capitalist society," where the means of production is the propriety of individuals in competition with each other. The production, as well, is carried on mainly for profit, not necessary for the benefit of others. This results in "an oligarchy of private capital," where the political parties of the members of the legislature are generously financed by such private capital.

> Moreover, … private capitalists inevitably control, directly or indirectly, the main sources of information (press, radio, education). It is thus extremely difficult … for the individual citizen to come to objective conclusions and to make intelligent use of his political rights.

Einstein therefore was convinced that "there is only one way to eliminate these grave evils, namely through the establishment of a socialist economy, accompanied by an educational system which would be orientated toward social goals." He was, however, not naïve about the possible problems for such a planned economy, a point he brought up at the end of the essay. Socialism must be instituted within a democratic system of government, to prevent a dictatorship; also, there must be a mechanism to prevent the rise of an all-powerful bureaucracy. Although he did not say it, he must have had the Soviet Union in mind. Thus only within a strong democratic society will true socialism work. Therefore, he seemed to say, America should have no fear of socialism, because it will not become Communism. And accordingly, the obsessive fear of Communism is emphatically irrational.

All this was in-theory. Still, he (along with Robeson) put their beliefs into practice. For example, they openly supported Henry A. Wallace, who was running for the Progressive Party in the 1948 Presidential election. Wallace's platform of universal government health care, an end to segregation, and full voting rights for blacks was closer to Einstein's social worldview than any other candidate. Wallace's progressivism was not socialism, but these were socialist policies.

Some Einstein biographers have called him naïve on social and political topics, and affirm that he should not have ventured into these non-scientific subjects. But anyone aware of the intertwining of politicians and capitalists in recent times, if not always, may rather be inclined to call him prophetic. For one, I do.

Indeed, in the last years of his life he devoted much work toward the idea of a world government. He saw this as the only means of avoiding a World War III, this one with nuclear weapons, which he believed would be the end of civilization. As he allegedly put it, his one case of dark humor I know of: World War IV will be fought with sticks, he said.

Another story during his American years that needs to be fleshed-out is his rescue of refugees (Jews and other dissenters) from Nazi Germany in the 1930s. Due to restrictive immigration rules, it was difficult for any German to enter the USA. So Einstein wrote affidavits for over 100 refugees trying to flee, and he paid some of the fees. At the time he wrote to his sister, "Miss Dukas and I run a kind of immigration office." Apparently, he paid a small fortune toward this effort.[132] The famous photographer, Philippe Halsman, said that Einstein's help probably saved his life. [133] Halsman went on to take one of the most famous portraits of Einstein, the one used for the *Time Magazine* cover in December 1999 when he was called the "Person of the Century."

All this shows a disconnect in his behavior. For those very close to him, he could be less than loveable – yet, in dealing with people from a distance, he proved to be a real *mensch*. (FYI: this Yiddish word – referring to one who is a very good person, who has integrity and should be admired – is the only word that really fits here.) For sure, there were fundamental contradictions in Einstein's personality. He was not religious in the formal or institutional sense of the term, but he had a strong sense of spirituality and moral duty. He was a loner, detached from others, but he had a deep concern with humanity as a whole and worked tirelessly on many social problems. In light of what we have seen of his life, it was not incongruous for him to make a declaration such as this: "Only a life lived for others is a life worth living." [134]

For his various do-gooder efforts, as seen in the previous chapter, J. Edgar Hoover and his FBI rewarded Einstein by opening a file on him. A cursory assessment of this file may be gleaned by looking at the 1940 document titled "Biographical Sketch." From the start and into the first few sentences there are five factual errors.[135] With this level of competence, the FBI pursued Einstein's social activities because they smacked of left-wingism – which, in the FBI's skewed view of the world, bordered on Communism. Einstein's opposition to McCarthy's shenanigans simply reinforced the FBI's case against him. What today we see as Einstein using his celebrity status to make a difference in the plight of the less fortunate in society, the FBI saw as a subversive activity. Paranoia can result in strange perceptions.

Begun in 1932, Einstein's FBI file grew by his death to about 1800 pages, a quest that was a total waste of time and money, for Einstein was no soviet spy, and he was not inventing some sort of "death ray," which the FBI believed. And they did not give up the chase until he was dead. The irony, absurdity, and even hilarity in all this is that there really was a soviet spy in the story. It was not Einstein, but an attractive woman friend of his, Margarita Konenkova, who he liaisoned with from 1935-1945. But she never pried any classified military secrets from him because he did not have any to divulge. Incredibly, none of this appears in the FBI Einstein file, for they were too stupid even to find this spy right under Einstein's nose, literally.[136] Folly and incompetence seen to sum up the affair (in both senses).

Einstein had no such secrets because he was considered a security risk from the time he immigrated to the USA in 1933. It is true that he was involved in the writing of a letter to President Roosevelt in August, 1939, that warned of the possible development of nuclear weapons by the Germans, and this may have played a role in the Manhattan Project that led to the creation of the first nuclear bombs dropped on Japan. But Einstein was not at Los Alamos, New Mexico with the scientists and engineers who put the bombs together from 1942 to the first test in 1945. He was not only banned from Los Alamo, but those working there were forbidden to consult with him.

Perhaps it is important to flesh-out this story a bit more. Here is an outline. The famous letter was initiated and drafted by several prominent scientists, most importantly, along with Einstein, was Leo Szilard, a Hungarian-American physicist. Szilard had been a student of Einstein and Planck in Berlin and moved to London in 1933 fleeing the Nazis, later living in the USA. Most historians now think that the letter probably was not very important; or, at least, not as important as a report from the British in September 1941, which contained scientific evidence that a bomb could be made in time to be used in the war, and this initiated the Manhattan project.[137] Beyond the letter, Einstein was not involved in either the technical or the scientific aspects of the making of the bombs.

Nonetheless, he was occasionally informed of how the development was going through his casual contact with scientists working on the project. At most, he was involved in related work for the Navy, but this had nothing to do with nuclear bombs. [138]

Nonetheless, the July 1, 1946 cover of *Time Magazine* depicted Einstein in the foreground and a nuclear blast behind him, with the equations $\mathbf{E = mc^2}$ in the mushroom cloud. All this is a visual howler in two ways. It is true that the little equation showed that if a very small amount of mass is completely converted into energy, that amount of energy is enormous, and this was the law of nature underlying the bombs. But, to get from this equation to a bomb (or also a nuclear reactor), required two components. One was nuclear physics, a new area of science discovered in the 20th century and a field to which Einstein made no contribution; he never published a paper in the subject. In fact, in the mid-1930s, when there was much talk in the popular press about atomic energy being generated by using his equation, he was asked about this possibility by a reporter. Einstein replied that he was convinced that it was almost impossible because it takes "a lot of energy to get any energy out of a molecule, and the rest is lost." He then made an analogy – that it was like trying to shoot birds in the dark in a land with few birds.[139] How wrong he was; of course, that was before there was experimental evidence for a nuclear chain reaction. The other necessary component was knowledge of the technology and engineering required to make an actual bomb or a reactor. Einstein had nothing to do with any of this.[140] In short, the identification of Einstein with the bomb is another of the countless myths about him.

Helen Dukas once recalled the day the first bomb was dropped on Japan. She heard it on the radio, and when she told Einstein what happened, he exclaimed: "Oy Veh!" This is another untranslatable Yiddish phrase for (sort of) "Oh No!" or "Oh my God!" Dukas, when telling this story, translated it as "How terrible."

Ultimately, looking at "Saint Einstein" in the context of his whole life – and especially in light of the real lives of many of those men and women of past ages who lived early years in debauchery only to morph into exemplary adults working selflessly for others, and who thus were lifted into sainthood – well, then maybe Einstein was not too far flung from being a sort of saint. In his later years, he certainly was treated as a saintly sage who made profound and quotable pronouncements on diverse topics, both scientific and non-scientific.

Einstein died in the hospital in Princeton on April 18, 1955 at 1:15 am of a ruptured aortic aneurysm. The night nurse said that he mumbled something in German right before he died, but she did not understand it since she did not know German. [141] Unfortunately, Dukas was not there to translate. I would give anything to know what those last words were.

The nurse also gathered up his belongings, parts of which were pages of loose papers consisting of mathematical expressions, which also made no sense to her. The pages were, we know, his last attempt in his quest toward a unified field theory. He was still plugging away to the very end. Does the man never stop?....

What are you missing?

The aim of this book is to tell the story of Einstein in a text that the reader will not only start, but may actually finish. Being a quick read for anyone, there are only two excuses why you stopped: you got bored, or you died.

If neither – that is, you did finish it – and I still have your interest in the topic, you may wish to read the next few paragraphs on other possible books.

I begin with the big ones, and work down. The most recent doorstopper is Walter Isaacson's outstanding *Einstein: His Life and Universe* (2007), which was a best seller. But I am reasonably certain that few of those buyers got very far into the over 600 pages before leaving it sitting unread on the night table, even though it left out technical and mathematical details in order to appeal to the general reader. (Only Einstein geeks like me read the whole thing, and with relish.) The same is probably true of Albrecht Fölsing's first-rate *Albert Einstein: A Biography* (1993/1997), a previous monster of a book on Einstein that is available to intelligent laypeople. Still accessible, exceedingly readable, and mostly reliable is Ronald W. Clark's *Einstein: The Life and Times* (1972), the first big-book on the genius.

The only other large-scale work on Einstein that I mentioned several times in my book (remember?) is Abraham Pais's *"Subtle is the Lord": The Science and the Life of Albert Einstein* (1982), a magisterial work of technical detail that remains the best biography of the scientist *as* a scientist. But this book scares off most readers from the start as they thumb through and see the many mysterious equations throughout.

There are many medium-sized books on Einstein, his life, or specific facets thereof. Some of these appear in my **Bibliography**. On the topic of very short books, as noted before, some good ones are out-of-print, but many are still available from used book dealers. Yet, it is only since the mid-1980s that previously unknown documents on Einstein's early life were made public, and that a revised image of him has emerged. It is this revision you have read in my little book – or, to be fair, my interpretation of this revision.

Bibliography

Auster, Simon. November 25, 2013. Telephone interview regarding his visit with Einstein in August, 1952, when he was a student at Yeshiva University.

Begley, Louis. 2014. "The Good Place in Vicious France," essay review of *Village of Secrets: Defying the Nazis in Vichy France*, by Caroline Moorhead (Harper, 2014), in *The New York Review of Books* (December 18, 2014), pp. 64 – 66.

Brian, Denis. 2005. *The Unexpected Einstein: The Real Man Behind the Icon*. Hoboken, NJ: John Wiley & Sons.

Calaprice, Alice (ed.). 2005. *The New Quotable Einstein*. Princeton: Princeton University Press.

Calaprice, Alice, & Daniel Kennefick, & Robert Schulmann (eds.). 2015. *An Einstein Encyclopedia*. Princeton: Princeton University Press.

Clark, Ronald W. 1972. *Einstein: The Life and Times*. New York: Avon Books. This is the paperback reprint of the first edition of 1971.

Einstein, Albert. 1949. *Autobiographical Notes*. Trans. and edited by Paul A. Schilpp. La Salle & Chicago: Open Court Publishing. This is the corrected version of the original 1947 German manuscript, first published in *Albert Einstein: Philosopher-Scientist*, two volumes. ed. by Paul A. Schilpp. New York: Harper & Row. Vol. I, 3-95. Einstein penned this very brief autobiography at the request of Paul Schilpp to introduce this two-volume collection of essays on Einstein by a range of 20th century scientists, philosophers, and others.

Einstein, Albert. 1950. *Out of My Later Years*. New York: Philosophical Library.

Einstein, Albert. 1954. *Ideas and Opinions*. Trans. Sonya Bergmann. New York: Bonanza Books.

Einstein, Albert. 1960. *Relativity: The Special and the General Theory*. 15th edition. Trans. Robert W. Lawson. London: Methuen & Co.

Einstein, Albert. 1974. *The Meaning of Relativity*, Fifth Edition. Princeton: Princeton University Press, 1974. This is a reprint of the first edition (1922) of the May, 1921 Lectures at Princeton University. It has the Appendix for the Second Edition (1945) and Appendix II: *Relativistic Theory of the Non-Symmetric Field* (1955), the latter (written with Bruria Kaufman) is the last scientific paper Einstein published.

Einstein, Albert. 1986. *Letters to Solovine*. Intro. by Maurice Solovine. Trans. Wade Baskin. New York: Philosophical Library.

Einstein Papers. 1987+. *The Collected Papers of Albert Einstein*. Princeton: Princeton University Press. This series began in 1987 and is an on-going project. At present, it is at volume 14(1925). See:
http://einsteinpapers.press.princeton.edu/papers

Eisinger, Josef. 2011. *Einstein on the Road*. New York: Prometheus Books.

Feuer, Lewis. 1982. *Einstein and the Generations of Science*. Second Edition. New Brunswick: Transaction Books.

Fölsing, Albrecht. 1997. *Albert Einstein: A Biography.* Trans. Ewald Osers. New York: Viking. Originally published in German 1993.

Frank, Philipp. 1947. *Einstein: His Life and Times.* Trans. George Rosen. Edited and revised by Shuichi Kusaka. New York: Alfred A. Knopf.

Gimbel, Steven. 2012. *Einstein's Jewish Science: Physics at the Intersection of Politics and Religion.* Baltimore: The John's Hopkins University Press. This is a non-racist, serious, and nuanced approach to this delicate subject.

Hillman, Bruce J. & Birgit Ertl-Wagner & Bernd C. Wagner. 2015. *The Man Who Stalked Einstein: How Nazi Scientist Philipp Lenard Changed the Course of History.* Guilford, Connecticut: L.P., an Imprint of Rowman & Littlefield.

Holton, Gerald. 1973. *Thematic Origins of Scientific Thought: Kepler to Einstein.* Cambridge, MA: Harvard University Press.

Holton, Gerald. 1996. *Einstein, History, and other Passions: The Rebellion against Science at the End of the Twentieth Century.* Reading: Addison-Wesley.

Isaacson, Walter. 2007. *Einstein: His Life and Universe.* New York: Simon & Schuster.

Jammer, Max. 1999. *Einstein and Religion: Physics and Theology.* Princeton: Princeton University Press. This is the most thorough study of this topic that I know of. He is especially good on the influence of Spinoza.

Jerome, Fred. 2002. *The Einstein File: J. Edgar Hoover's Secret War Against the World's Most Famous Scientist.* New York: St. Martin's Press.

Jerome, Fred. 2009. *Einstein on Israel and Zionism: His Provocative Ideas about the Middle East.* New York: St. Martin's Press.

Jerome, Fred & Roger Taylor. 2005. *Einstein on Race & Racism.* New Brunswick, NJ & London: Rutgers University Press.

Kuhn, Thomas S. 1987. *Black-Body Theory and the Quantum Discontinuity, 1894-1912.* Chicago: University of Chicago Press. This is a reprint, with a new Afterword, of the 1978 edition.

Marić, Mileva. 2003. *In Albert's Shadow: The Life and Letters of Mileva Marić.* ed. Milan Popović. Trans. Boško Milosavljević & Branimir Živojinović. Baltimore: The Johns Hopkins University Press. The originals are in German and Serbian.

Martinez, Alberto. 2005. "Handling Evidence in History: The Case of Einstein's Wife," *School Science Review*, 86 (March), 49-56.

Miller, Arthur I. 2001. *Einstein, Picasso: Space, Time, and the Beauty That Causes Havoc.* New York, Basic Books.

Neffe, Jürgen. 2007. *Einstein: A Biography.* Trans. Shelley Frisch. New York: Farrar, Straus, & Giroux. Originally published in German in 2005.

Nussbaumer, Harry. 2014. "Einstein's Conversion from his Static to an Expanding Universe," *The European Physical Journal: Historical Perspectives on Contemporary Physics*, Vol. 39, No.1 (February, 2014), 37-62.

Pais, Abraham. 1982. *"Subtle is the Lord": The Science and the Life of Albert Einstein.* New York: Oxford University Press.

Parker, Barry. 2003. *Einstein: The Passions of a Scientist.* Amherst, NY: Prometheus Books.

Renn, Jürgen (ed.). 2005. *Albert Einstein: Chief Engineer of the Universe. Vol. I. Einstein's Life and Work in Context; Vol. II. One Hundred Authors for Einstein; Vol. III. Documents of a Life's Pathway.* Weinheim: Wiley-Vch Verlag.

Rosten, Leo. 1970. *The Joys of Yiddish.* New York: McGraw Hill, Pocket Books.

Sayen, Jamie. 1985. *Einstein in America: The Scientist's Conscience in the Age of Hitler and Hiroshima.* New York: Crown Publishers.

Schweber, Silvan S. 2008. *Einstein and Oppenheimer: The Meaning of Genius.* Cambridge, Mass.: Harvard University Press.

Solovine, Maurice. *See:* Einstein, 1986.

Staley, Richard. 2008. *Einstein's Generation: the Origins of the Relativity Revolution.* Chicago: University of Chicago Press.

Stern, Fritz. 1999. *Einstein's German World.* Princeton: Princeton University Press. See especially Chapter 3, "Together and Apart: Fritz Haber and Albert Einstein."

Topper, David R. 2007. *Quirky Sides of Scientist: True Tales of Ingenuity and Error from Physics and Astronomy.* New York: Springer.

Topper, David R. 2013. *How Einstein Created Relativity out of Physics and Astronomy.* New York: Springer.

Topper, David R, 2014a. *Idolatry & Infinity: Of Art, Math, & God.* Boca Raton, Florida: BrownWalker Press.

Topper, David R. 2014b. "Einstein's Blunder," *The New York Review of Books* (May 8, 2014), p. 57. A letter to the editor.

Topper, David R. & Dwight Vincent, 2007. "Einstein's 1934 Two-Blackboard Derivation of Energy-Mass Equivalence," *American Journal of Physics*, 75 (November), 978- 983.

Vallentin, Antonina. 1954. *The Drama of Albert Einstein.* Trans. Moura Budberg. Garden City, New York: Doubleday.

Wazeck, Milena. 2014. *Einstein's Opponents: The Pubic Controversy about the Theory of Relativity in the 1920s.* Trans. Geoffrey S. Koby. Cambridge, UK: Cambridge University Press. The original German edition was published in 2009.

Winteler-Einstein, Maja. 1987. "Albert Einstein – A Biographical Sketch (excerpt)," in *Einstein Papers*, 1987, Vol. 1, pp. xv – xxiii. His sister wrote this sketch in 1924.

EndNotes

[1] Two earlier pictures exist: at age three in a formal outfit, and a studio shot at age five with his sister.

[2] Brian, 2005, p. 8. Brian, Barbara Wolff, and I are the only authors I know of who have noticed this.

[3] Brian, 2005, pp. 104-105.

[4] Brian, 2005, p. 6.

[5] Winteler-Einstein, 1987, p. xxii.

[6] Einstein, 1949, pp. 4–5.

[7] Topper, 2013, p. 11.

[8] Winteler-Einstein, 1987, pp. xxi-xxii.

[9] Einstein, 1949, p. 9.

[10] Brian, 2005, p. 12.

[11] Holton, 1973, pp. 370-373.

[12] Holton, 1996, pp. 390-391.

[13] Miller, 2001, p. 48, has also noticed this.

[14] Miller, 2001, p. 186.

[15] *Einstein Papers,* Vol. 1, Doc. 18.

[16] For example, Neffe, 2007, p. 88.

[17] *Einstein Papers,* Vol. 1, Doc. 34.

[18] *Einstein Papers,* Vol. 1, Doc. 36.

[19] *Einstein Papers,* Vol. 1, Doc. 72.

[20] *Einstein Papers*, Vol. 1, Doc. 58.

[21] *Einstein Papers*, Vol. 1, Doc. 50.

[22] Parker, 2003, p. 46.

[23] *Einstein Papers*, Vol. 1, Doc. 75.

[24] Marić, 2003, p. 79.

[25] Brian, 2005, Chapter 4 is good summary.

[26] Einstein, 1986.

[27] *Einstein Papers*, Vol. 1, Doc. 131.

[28] Einstein, of course, was a non-practicing Jew, and she Serbian Orthodox.

[29] Martinez, 2005, dispels the rumour that Mileva was instrumental in the creation of relativity theory.

[30] *Einstein Papers*, Vol. 5, Doc. 374.

[31] Feuer, 1982, pp. xii-xix; Topper, 2013, pp. 74-75.

[32] Topper, 2013, pp. 70-73.

[33] Parker, 2003, p, 112.

[34] *Einstein Papers*, Vol.8, Doc. 22 (July 18, 1914).

[35] Vallentin, 1954, pp. 86-90.

[36] *Einstein Papers*, Vol. 8, Doc. 34a (in Vol. 10, p. 12).

[37] *Einstein Papers*, Vol. 10, Docs. 154, 165, & 166.

[38] Frank, 1947, pp. 177-178.

[39] Frank, 1947, p. 226. The cottage was at Caputh, about 20 miles southwest of Berlin, near Potsdam.

[40] Topper, 2013, p. 216; Neffe, 2007, pp. 199.

[41] Vallentin, 1954, p. 240.

[42] Calaprice, 2005, p. 73.

[43] *Einstein Papers*, Vol. 1, Doc. 111.

[44] On this debate see, Topper, 2013, pp. 74-75.

[45] Calaprice, 2005, pp. 275-277.

[46] *Einstein Papers*, Vol. 8A, Doc. 257. My translation.

[47] Solovine, 1986, pp. 8-9.

[48] Jammer, 1999, especially pp. 146-149.

[49] Calaprice, 2005, p. 197.

[50] Calaprice, 2005, p. 196.

[51] Auster, 2013.

[52] Stern, 1999, p. 80.

[53] Neffe, 2007, pp. 317-318.

[54] Einstein, 1954, p. 188.

[55] Stern, 1999, p. 74.

[56] *Einstein Papers*, Vol. 8, Doc. 34a (in Vol. 10, p. 12).

[57] Stern, 1999, p. 140.

[58] Topper, 2013, p. 123; see more recent work by Hillman, et al., 2015, pp. 57-58 and especially Wazeck, 2014, pp. 3-4 & 148-154.

[59] Gimbel, 2012, especially Chapter 4.

[60] Renn, 2005, II, p. 224.

[61] A recent book on this is Hillman, et al., 2015.

[62] Renn, 2005, II, pp. 310-313.

[63] Jerome, 2009, p. 102.

[64] Einstein, 1986, p. 41 (letter of March 8, 1921).

[65] Einstein, 1974.

[66] *Einstein Papers*, Vol. 7, Doc 50.

[67] Gimbel, 2012, p. 132.

[68] Stern, 1999, p. 194n.

[69] Einstein, 1986, p. 57.

[70] Einstein, 1986, p. 59.

[71] Clark, 1972, p. 477.

[72] *Einstein Papers*, Vol. 13, Intro. p. lxvi.

[73] Clark, 1972, pp. 477-479.

[74] "Waffle" is a good British-Canadian expression, meaning to vacillate back and forth.

[75] Calaprice, 2005, p. 129.

[76] Calaprice, 2005, p. 132.

[77] Topper, 2013, p. 124.

[78] Topper, 2013, p. 192.

[79] Stern, 1999, p. 153.

[80] Begley, 2014, p. 64.

[81] Calaprice, 2005, p. 162.

[82] Stern, 1999, p. 162.

[83] http://www.conservapedia.com/Main_Page

[84] *Einstein Papers*, Vol. 5, Doc. 445.

[85] Eisinger, 2011, p. 145.

[86] Einstein, 1950, pp. 126-129.

[87] Jerome and Taylor, 2005, pp. 88-92 & 142.

[88] Chaim Weismann, who took Einstein on the tour of the USA, was the first.

[89] Jerome, 2009, p. 76.

[90] Jerome, 2009, pp. 99-100.

[91] Jerome, 2009, p. 112.

[92] Jerome, 2009, p. 157.

[93] Einstein, 1974. See pages 113 & 130 for Einstein's 1945 comments on the three possible cosmic spaces: positive curvature, negative curvature, or flat.

[94] Topper, 2014a, pp. 79-83 for a brief summary.

[95] Topper, 2013, Chapter 2.

[96] Staley, 2008, is an excellent book on this topic.

[97] *Einstein Papers*, Vol. 1, Docs. 27 & 28.

[98] Parker, 2003, pp. 102-103, 107-109, & 112-113.

[99] Kuhn, 1987.

[100] Along with physicist Paul Ehrenfest, who later became a close friend. In 1905-1907 each was working independently.

[101] *Einstein Papers*, Vol. 5, Doc. 445.

[102] Topper, 2013, p. 40. He used this word in a letter of 1947.

[103] The first electric elevator appeared in 1857. Topper, 2013, p. 89.

[104] For more details of this story, see Topper, pp. 89-91.

[105] *Einstein Papers*, Vol. 1, Doc. 131, December 1901.

[106] Einstein, 1954, pp. 289-290. From the Glasgow lecture of 1933.

[107] Topper, 2013, p. 116.

[108] Einstein, 1960.

[109] Pais, 1982, p. 182.

[110] Topper, 2013, p. 108.

[111] Topper, 2013, Chapter 22.

[112] Topper, 2013, pp, 180-181.

[113] Topper, 2013, p. 168 & 168n14.

[114] Nussbaumer, 2014, presents strong evidence for revising the conventional thesis that Einstein's change of heart took place at Caltech.

[115] Topper, 2013, pp. 173-174. Nussbaumer makes a further assertion that Einstein probably had little contact with Hubble at Caltech, but was exposed to the astronomical data through Richard C. Tolman, the physicist and chemist, who he worked with during that trip.

[116] Topper, 2014b. Contrary to a recent assertion that this quote is a myth, I present strong evidence that the quote is indeed true. For another reference to this story, see Topper (2013), Footnote #1 on p. 165. In both cases I cite a confirmation by the cosmologist, Ralph A. Alpher, in an email of April 2, 1998, who was present at the meeting when Einstein made the remark.

[117] Topper, 2013, pp. 132-134.

[118] Topper, 2013, p. 218.

[119] Topper, 2013, p. 223.

[120] Vallentin, 1954, p. 259.

[121] Schweber, 2008, p. 35.

[122] A simple calculation is in Topper, 2007, p. 11. A discussion of this experiment is on pp. 10-12.

[123] See my book, Topper, 2007, Chapter 13. As far as I know, this explanation in unique with me.

[124] Topper, 2013, both quotes from p. 102.

[125] Topper, 2013, quoted on p. 203. Pais' italics.

[126] Calaprice, et al. 2015, pp. 178-179.

[127] Vallentin, 1954, pp. 154-155; Jammer, 1999, *passim*.

[128] Topper, 2013, p. 212.

[129] Neffe, 2007, pp 102-103; Vallentin, 1954, pp. 141-144.

[130] Brian, 2005, p. 51.

[131] Einstein, 1954, pp.151-158.

[132] Neffe, 2007, p. 377; Frank, 1947, pp. 275-278; Sayen, 1985, Chapter 6, esp. pp. 112-116.

[133] Brian, 2005, p. 154.

[134] Vallentin, 1954, p. 93.

[135] Jerome, 2002, illustration between p. 170 & 171. The five errors are: (1) in the USA Einstein was at the Institute for Advanced Study, not Princeton University; (2) Mileva's last name was Marić, not Maree; (3) he married Mileva in 1903, not 1901; (4) he married Elsa in 1919, not 1917; and (5) he and Mileva had two children, not one (of course, few knew about the third child, Lieserl, at this time, i.e., 1940).

[136] Brian, 2005, Chapter 9.

[137] Fölsing, 1997, pp. 713-714.

[138] Schweber, 2008, pp. 42-62.

[139] Topper and Vincent, 2007, p. 979.

[140] Topper, 2013, p. 220.

[141] Topper, 2013, p. 219.

www.ingramcontent.com/pod-product-compliance
Lightning Source LLC
Chambersburg PA
CBHW071454160426
43195CB00013B/2105